Shetland Breeds . . .

Ancient, Endangered & Adaptable

Where besides Shetland do so many
indigenous and endangered farm animals thrive?

A compendium with principal essays by

Andro Linklater

Valerie Russell

James R. Nicolson

Ronnie Eunson

Nancy Kohlberg

Lawrence Alderson

Richard H. L. Lutwyche

Edited by Nancy Kohlberg and Philip Kopper

Shetland Breeds

'Little Animals... Very Full of Spirit'

Ancient, Endangered & Adaptable

Library of Congress Cataloging-in-Publication Data
Shetland breeds, 'little animals… very full of spirit' : Ancient, Endangered & Adaptable / byAndro Linklater. . . [et al].
Nancy Kohlberg and Philip Kopper, eds.
p. cm
Includes bibliographical references (p.)
ISBN 1-889274-10-0
1. Livestock breeds—Scotland—Shetland—History 2. Livestock—Scotland—Shetland—History. I. Linklater, Andro
SF53 .S54 2001
636.08'2'0941135—dc2100-066873

Front cover: A Shetland ram, patriarch of the Shetland sheep, one of several breeds of farm animals indigenous to the most remote of the British Isles, the Shetland Islands, called in the past Zetland. Painting by Robin Hill.

Back cover: Scalloway Castle, the ruin that remains in the new millennium much as it appears here, circa 1848, commanding land and harbor on the west coast of Shetland's Mainland, where Shetland cattle graze, a Shetland pony wanders and crofters watch a visiting artist at his easel. The watercolor, by John Christian Schetky (1778–1874), graces the travelogue *Sketches and Notes of a Cruise in Scotch Waters on Board His Grace the Duke of Rutland's Yacht Resolution* (London 1849).

Endpapers: Shetland breeders prepare their animals for judging in the annual Shetland County Agricultural Show at the turn of the last century.

Posterity Press, Inc.
PO Box 71081
Chevy Chase, MD 20813
www.PosterityPress.com

No man is an island, entire of itself; . . .

—JOHN DONNE
Devotions XVII
1624

Zetland is as yet a little world by itself, . . .

—SIR WALTER SCOTT
The Pirate
1821

Contents

Illustrations

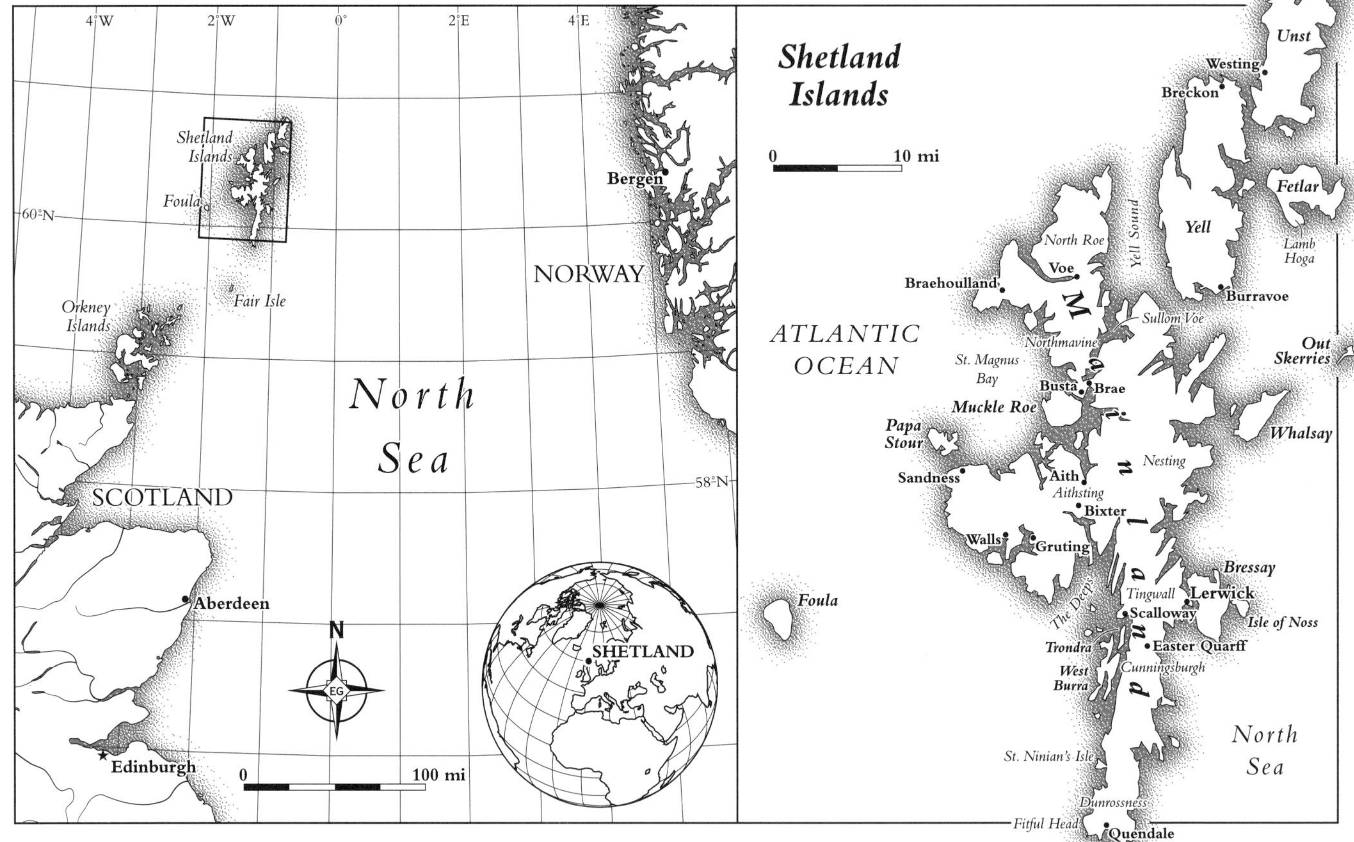
4°W
2°W
0°
2°E
4°E
60°N
58°N
Shetland Islands
Foula
Fair Isle
Orkney Islands
NORWAY
Bergen
North Sea
SCOTLAND
Aberdeen
Edinburgh
N
EG
0
100 mi
SHETLAND
Shetland Islands
0
10 mi
Herma Ness
Unst
Westing
Breckon
Fetlar
Yell
Yell Sound
North Roe
Lamb Hoga
Voe
Braehoulland
Burravoe
Sullom Voe
Mainland
ATLANTIC OCEAN
Northmavine
Out Skerries
St. Magnus Bay
Busta
Brae
Muckle Roe
Papa Stour
Whalsay
Nesting
Sandness
Aith
Aithsting
Bixter
Walls
Gruting
Bressay
Tingwall
Lerwick
Foula
The Deeps
Scalloway
Isle of Noss
Trondra
Easter Quarff
West Burra
Cunningsburgh
North Sea
St. Ninian's Isle
Dunrossness
Fitful Head
Quendale
Jarlshof
Sumburgh Head

Preface

Norman Lamont

The Rt. Hon. Lord Lamont of Lerwick

I am delighted to introduce *Shetland Breeds, 'Little Animals. . . Very Full of Spirit.'* This beautiful book will be a real pleasure for people to read and to possess. It is both intriguing and instructive as it describes as wide an array of indigenous breeds of domestic animals as may be found in any locality on earth: Shetland Pony, Shetland Sheep, Shetland Cattle, Shetland Goose, and more.

Islands often produce subspecies of the main species of wild birds and animals. Thus the Shetland Wren, like the St. Kilda Wren and Hebridean Wren, is slightly different from that found on the mainland of Britain, and the Shetland Starling, *Sturnus vulgaris zetlandicus,* is different from the common *Sturnus vulgaris.* The effects of geographical isolation dividing species into subspecies can be astonishingly far-reaching.

Shetland not only has its own field mouse, but so remarkably do individual islands in Shetland, ergo the Fair Isle Field Mouse and Foula Field Mouse, both of them properly and scientifically classified.

I had not appreciated how this phenomenon of local specialization of wild fauna also applies to domestic animals and birds: ducks, geese, sheep, ponies and kine. But upon reflection it is obvious. I have to confess that fiercely patriotic and proud Shetlander though I am, I was not aware of Shetland Cattle, and I had never seen a Shetland Goose. As a child in Shetland I was obsessed with rediscovering the vanished Great Auk, and now I have an alternative dream of finding the missing Shetland Pig. But did it ever exist? That is one of the interesting questions explored in this book.

As one chapter explains, today there are some 15,000 Shetland Ponies in all of Britain (only a fraction in Shetland itself) and 75,000 worldwide. Of course, away from their native habitat the animals of Shetland do not remain the same; they change as they adapt to a new environment. Even in Shetland itself there is an understandable tendency to import other species for cross-breeding to see if animals can grow more quickly, as in the case of cattle, or grow larger, as in the case of Shetland Sheep when crossed with Cheviot and Blackface. It is surely important to retain a population of purebred animals in their own environment to whom breeders can return for an infusion of true blood.

Today we hear much about the need for biodiversity, and that admirable concept should include conservation of local subspecies, which can only be done in their native environment. We owe a great debt of gratitude to the individuals and societies who have worked to keep Shetland breeds alive. The Shetland Cattle Herd Book Society, founded in 1912, has played a vital role in bringing the "Shetland Coo" back from the brink of extinction. Now it has been discovered that her milk, when she feeds on native grass, has distinctive anti-carcinogenic qualities, another proof of how right it was to save these animals.

This book itself is almost indigenous. Although published by an American house, all but one of its authors are Britons, a plurality of them Shetlanders. They cover their topics with admirable expertise and bring to bear more than the sum of their separate fields of knowledge and experience. Their essays focus on the animals, yet the book reveals a larger subject: its principal protagonist is Shetland itself, and its heroes quite rightly are the people of those islands.

LONDON

Foreword

John Scott

Lord Lieutenant of Shetland

This timely book comes at a point when the issues raised by the authors are being widely debated. All Shetlanders live with their past and the reader will find the present addressed in the perspective of our history. Animals arrived in these islands less than 12,000 years ago, in the wake of the retreating ice. Modern geneticists have shown how a small sample of a breed carries only a fraction of the genetic inheritance of the whole breed and will therefore quickly show variations from the parent stock; thus arose these Shetland breeds.

In this book the authors tell of their breeds—Valerie Russell and the Shetland pony, Mary Isbister and the Shetland duck, Ronnie Eunson and Shetland cattle, Agnes Leask and Shetland sheep, and more. The human side of the story is retailed by James R. Nicolson from his encyclopedic knowledge of the islands and by Andro Linklater in his history, while Lawrence Alderson explains island genetics and Richard Lutwyche defines the phantom grice.

The reader should envisage the nature of the place that is Shetland: sixty degrees north in the latitude of Cape Farewell in Greenland and Labrador in Canada, swept by winter storms in a treeless landscape on the very edge of Europe where the continental shelf suddenly plunges to 2,000 fathoms. No wonder the livestock had to be adaptable; no wonder Shetlanders are too.

The isolation of island breeds is of course a classic situation for the study of genetic variation. Professor Sam Berry's studies of field mice in the 1960s led the way in the spirit of scientific discovery so profoundly influenced by Darwin. Scientific research has its lighter moments. How we used to look forward to Sam's visits to the farm here on winter days when he pounced on the mice as we took in the sheaves of oats for threshing. It was fun, but that research illuminates what used to be as well as what may come after.

The comprehension of the importance of Shetland's genetic pool and the way our domestic breeds have been developed and adapted acquires greater cogency in the face of climate change and access to new bloodlines. In itself however it is a fine and often heartwarming story of Shetlanders and their stock on which the rural economy of the islands depends.

Over the last three decades Shetland has once again become, as it was in former days, the crossroads of the North Sea. The islands stand poised at the center of oil development in the northern North Sea and the emergent Atlantic fields. Exciting challenges confront us. We have to face the dangers posed to our pristine natural environment from the mechanics of offshore drilling. Our fishing industry has had remarkable success but now must face the reality of dwindling stocks. Agriculture also depends on a healthy environment—the natural grazings of heather hill and clean maritime grassland. Tourism is largely fueled by interest in wildlife—everything from otters to spectacular seabird colonies and other fauna, which all must be protected.

All of this, and the culture which it has engendered, depends on the sustainability of our local industries, but we, like other island dwellers globally, find ourselves in the van of awareness of and susceptibility to climate change. We see the sea level rising, and the ocean currents waver. Survival may become an issue for us rather sooner than for the inhabitants of continents who think themselves immune from the great forces of nature. It would be a sad irony indeed were this book to become an epitaph for Shetland's native breeds.

As I write, the brisk airs of April scour the land, a hint of fugitive green warms the fields, and the first song of the lark signals the resurgence of life, to be followed by the great annual affirmation of the lambing. The publication of this book by Posterity Press is an admirable project. It brings to the reader a fascinating story from these Isles of Shetland.

BRESSAY

SHETLAND OVERVIEW

HISTORY IN A HARSH HABITAT

One of Shetland's many rural townships, Fogrigarth in Aithsting, as it appeared a lifetime ago in the 1930s. The summer sun still shines bright on the hamlet of houses, byres, and enclosed gardens, when grain stands in sheaves in the mown fields.

Eden with Weather

A Short History of This Valiant Place

Andro Linklater

It was an afternoon in late May, and a blustering westerly wind was building the North Atlantic into vast, slow rollers. As each one struck the rocks at the foot of the cliffs, it exploded in a million molecules of spray which drifted like mist and slowly dissolved until the destruction of the next wave thickened the air again. Above the mist appeared a flock of puffins returning to their nests from fishing grounds out in the shallow seas beyond the island of Papa Stour.

With dumpy black bodies and red bulbous beaks, the puffins looked like clowns. In the air their stubby wings beat so fast, they were only a blur, but under the water they would have been graceful, flying sinuously after pencil-length sand eels, more fish than fowl, until they filled bellies and beaks with the silver eels. At some imperceptible signal of their communal mind, they had bobbed to the surface, and after a moment to gather themselves together, taken off and headed for home, the colony of nests they had burrowed out on top of the cliffs of Unst.

Slower or greedier than the others, one puffin let a gap of fifty yards develop between itself and the rest of the flock. That was an error and at 60 degrees north latitude, the margin for error is very slim. Drifting heavily on the updraft above the cliffs were three or four great skuas. In Shetland, these birds are called bonxies. It is a good name, as hard and vicious as the winged mugger it describes. Armed with a four-foot wingspan and a heavy, hooked beak, the bonxie lives by ambushing the other seabirds on the cliffs.

The tail-ender was less than a hundred yards from safety when the first bonxie swooped down on it. Its target swerved violently, straightened its flight, and was immediately blanketed by the wings of a second dive-bomber. Flinging itself upwards the puffin narrowly escaped being beaten into the sea, but its forward motion had stopped. Its flailing wings tore at the air to avoid a stall. As it hung there the first bonxie gripped it by the tail and tipped it into the waves, then settled in the water beside it with wings raised, screaming ferociously. Accepting defeat, the little puffin opened its comical beak and disgorged the eels it had caught. Scrabbling and squawking in the waves, the predators devoured a meal intended for the puffin's chicks.

The murder happened quickly, but across the years it remains an indelible memory. Perhaps there was not enough food to satisfy them or perhaps they were in a vicious mood, but one of the bonxies suddenly pecked at the puffin's head. The little bird flapped its wings trying to escape. There was another peck, then the second bonxie joined in. And after a time they flew away. The puffin was lying rather than bobbing on the water, but it must still have been alive, for occasionally its wings fluttered. The movement attracted a third bonxie which came down to investigate. Almost casually, it began pecking at the battered body. The fluttering stopped. Another jab of the hooked beak, then, either bored or convinced that it would not make a meal, the bonxie took advantage of a peaked wave and lifted off into the wind. The dead puffin drifted limply, a little black dot in the vast, crumpled water.

Set in the midst of the ocean, with 3,000 miles of Atlantic to the west, and 200 miles of North Sea to the east, Unst lies at the top of the 60-mile-long string of a hundred or more islands known collectively as Shetland, midway between the tip of Britain and Norway. There are few places in the northern hemisphere more isolated. Just fifteen of the islands are inhabited, the largest being called the Mainland, the smallest the tiny square-mile cluster of the Out Skerries. Most lie on the Arctic side of 60 degrees north, a line of latitude which runs past Bergen in Norway, then St. Petersburg, Russia, and through the frozen tundra in Siberia, to touch Alaska, the ice floes of Hudson Bay and Cape Farewell in Greenland before completing the circle at the southern tip of the Mainland. By rights, then, the islands' climate should be sub-Arctic—ice and snow melted by a brief summer sun—but a freak of geography, the Gulf Stream, that massive river of warm water surging up from the Gulf of Mexico and all the way across the North Atlantic, drives away the glaciers and bergs. A Shetland winter ranges from raw to fierce and summer can be as memorable for its caressing warmth as for its endless drizzle.

"I know bonxies are part of the natural order of things," said Tommy Isbister when I described the long-ago incident, "but seeing them attack puffins and tirricks [terns] still makes me angry. Life's hard enough without having a bonxie stealing the chicks' supper." With his wife, Mary, Tommy farms 55 acres of heather and rough pasture on Trondra, an island linked to Mainland by a one-lane bridge, and living at the very edge of the arable world, Shetland farmers are experts on the hardness of life. For generations they have learned to extract a living from an environment whose demands are literally unique.

As true for people as it is for birds, nature's extremes leave no leeway to deal with the unexpected hazard. The natural world is starker at these latitudes. The storms carry a heavier punch, the windchill cuts deeper. In winter, night swamps the sun and gales and blizzards often build to hurricane force, with a gust of 200 miles an hour on record. In summer, day squeezes out night and the dark at midnight is no more than a pearly grayness called "the summer dim." Through the long hours of light, the wind, sun and rain revolve in a turbulent cycle which can suddenly be broken by days of still, halcyon sunshine, known in the wary Shetland phrase as "days atween wadders" (days between weather).

I went to see Tommy Isbister on a tumultuous winter's morning when the wind tumbled every loose object around a muddy farmyard in a wild, witch's dance. Sensibly he had retreated inside, and was serenely sighting down the graceful curve of the wooden boat which he was constructing in the peace of a well-lit workspace. Sparely built with thin silver hair, he smiled with an ease that crinkled his blue eyes as though he had not a care in the world. Yet a farmer further south, accustomed to raising heavily built Holstein cattle or big-boned Cheviot sheep on rich grassland, would have been in despair at the prospect of keeping livestock in his fields, and especially in an icy blast that cut through heavy-weather clothing as though it were fine cotton.

The Isbisters' animals were different: small black cattle, diminutive, dark-fleeced sheep, and shaggy little ponies, all with their tails turned to the wind, but busily grazing on the winter pasture, oblivious to the cold. In the farmyard, glossy black ducks and grey geese rootled for leftover grain in the lee of the barn, seemingly immune to the gusts which ruffled their feathers. "If they were the usual commercial breeds, they would have be kept inside in this weather and they would need feeding all through the winter," Tommy observed. "But these are Shetland animals, and they're adapted to conditions like this."

In fact what his fields contained was a history lesson. Until the nineteenth century, Shetland farmers existed at almost subsistence level. Most years grain had to be imported, and the farmers relied on fishing and the sale of knitwear to fill the gap. To survive, they evolved breeds of animal found nowhere else in the world, which were able to thrive on poor nutrition and in ferocious weather. Their most obvious characteristics are small size and warm coats, but from close up another equally important trait becomes apparent. The *Shetland Flock Book* specifies that the eyes of a Shetland sheep should have a "full, bright and active look." The Isbisters' sheep clearly conformed, but so too did their bright-eyed, inquisitive ponies and their quick-moving, responsive cattle.

As long ago as 1808, John Shirreff, a visitor to Shetland, remarked on the size of the islands' domestic animals, "the smallest of any in the British empire," he declared. He also commented on that intangible but immediately recognizable aspect, their character. The cattle were hardier than "animals from warmer regions," the dogs were scruffier, the pigs more savage, and the horses feistier. "These little animals are very

Mary Bowie sits her son Tom on a prized pony at the family seat, Park Hall near Bixter circa 1910. Her husband, Dr. James Bowie, championed Shetland's breeds early in the 20th century, and their sons will follow his example: Stanley would become a geologist in England yet still publish monographs on Shetland breeds. Hugh who stayed home, a life-long crofter, was one of two islanders who kept the last small herds of purebred Shetland Cattle until others joined their cause and rescued the breed from extinction.

full of spirit," he told his readers, "and can bear fatigue much better than larger horses."

Traditionally farming was a part-time occupation, shared with fishing, knitting or some other craft. Even today only the larger farms earn enough to carry a family through the long, unproductive winters. The boat taking shape under Tommy's hands was his winter task. It was a 26-foot-long sixern, the traditional, open, lapstrake boat which Shetlanders used for fishing almost until the twentieth century. It took its name from the six men it required, each pulling a single oar, to propel the boat out to the fishing grounds ten or twenty miles offshore. Like the livestock in the fields, its shape was perfectly adapted to its environment, the crested, curling seas that whip up when coastal tides cut across broad ocean rollers. Tall and pointed at each end, the sixern grew broad and shallow in the middle and, although barely a quarter the length, its proportions exactly resembled a Viking longship. The similarity was no coincidence for it was in those craft that Shetland's most famous settlers arrived more than a thousand years ago.

According to *Egil's Saga,* one of the great Icelandic tales which tell the story of the Vikings, the first Norsemen came to Shetland seeking land. Like the nineteenth-century pioneers who headed west across the North American plains, they sailed from Scandinavia across the North Sea towards the setting sun in search of a place to cultivate crops and raise animals. Their covered wagons were the wonderfully flexible longships which twisted and slid rather than slammed through the waves, and just as the legend of the American West was to glorify the gunfighter and ignore the cowhand, so the sagas are filled with the deeds of warriors rather than of farmers. In each case, the reality was that herding and ploughing left a deeper mark than shedding blood.

The Norse pioneers made new homes wherever they could find room, from Iceland in the north to Ireland in the south, and above all in Orkney and Shetland, the two groups of islands which all voyagers encountered as they sailed north of the British landmass. Although the earliest of these settlers would have arrived before 800 A.D., they were not the first Shetlanders. According to the twelfth-century *Historia Norvegiae,* the Vikings found the islands inhabited by Picts and Papae: "the Picts little exceeded Pigmies in stature," the history records; "they did marvels in the morning and the evening, in building [walled] towns, but at midday they entirely lost their strength and lurked through fear in little underground houses." The

Papae, the text continues, "have been named from their white robes, which they wore like priests."

However mythical they might sound, solid evidence for the existence of both Picts and Papae can still be found, and for even earlier inhabitants. At Jarlshof, a low mound within sight of Shetland's main airport at Sumburgh, are the remains of houses and farms up to 4,000 years old. The earliest buildings are stone-lined chambers dug into the ground which date back to perhaps 2000 B.C., or the late Stone Age. Archaeological evidence shows that the inhabitants kept sheep and cattle and grew a primitive barley, thus demonstrating that farming here is as old as human habitation. Since these first Shetlanders also built large semi-circular or heel-shaped religious buildings, similar to temple sites otherwise found only in the Mediterranean, it has been suggested that they might have come from far away in the sunlit south.

The mystery of their origins died with them, however, and sand blown in from nearby dunes eventually covered over their farms, so that when later Bronze Age inhabitants arrived they built their homes on top of the first chambers. These new buildings were circular huts, like African *rondavels,* with a central fire. Almost certainly their owners were also responsible for constructing the tall, bell-shaped towers called *brochs* which still survive all over Shetland. Almost 40 feet high and usually sited close to the sea, the brochs must have served as lookout points, and possibly have sheltered a small flock of animals against marauders. With walls up to 20 feet thick, and constructed without mortar, they are mighty defensive works, yet in the end they apparently failed to protect the people who made them. By about 350 A.D., the broch builders had been displaced by a third wave of incomers. Their homes, still circular, also had a fireplace at the center but now seven or eight narrow bedrooms radiated out like spokes from the warmth, a shape which led them to be known as "wheel-houses."

The people who built the wheel-houses were the Picts. It is impossible to know whether they were as small or as alternately energetic and apathetic as the Norse history claims, but they were certainly builders. Entire islands were divided by massive turf and stone walls or dykes, known even today as "pickie dykes." The Picts were converted to Christianity by white-robed monks who could well have been known as Papae since their centers of worship were commemorated by the name "Papa" as in the island of Papa Stour. That much at least of the history was accurate but, apart from some unin-

Donald McAdie, who tended the horses at Noss for the Marquis of Londonderry, hefts one of his charges in 1901. It must be near autumn, to judge by the heap of peats piled near the croft house door. A half-century earlier, the fifth Marquis pioneered the use of sturdy Shetland ponies in his coal mines in England—after Parliament passed laws barring women and children from the pits.

telligible writing, the only records that the Picts left of themselves was their art. In 1958, a Pictish church was excavated on St. Ninian's Isle and yielded one of the great treasure troves of modern times—a hoard of intricately decorated silver ornaments including bowls, brooches and a Communion spoon. At Papil on the island of West Burra, a remarkable carved stone found on the site of another church suggests something of their fantastic imagination. The Christian symbols of cross and crosier occupy the upper half of the carving, but below it stands a beast with the body of a dog and the head of a boar, and at the foot two men with heads like herons have transfixed a human skull with their long beaks. It is an image which triggers many ideas. Between the human and natural worlds must have existed an intense, almost totemic relationship. And surely the Picts also understood about the unexpected blow of fate, the bonxie attack.

The earliest representation of Shetland ponies comes, not surprisingly, in a Pictish carving, one showing a line of monks on horseback, and the evidence of archaeology shows that the Picts also bred cows and sheep. Yet for all their art and agriculture, the arrival of the vigorous, assertive Vikings in the eighth century obliterated the Picts so completely that nothing more is known of them than can be gleaned from these remains of their rich culture.

The remains at Jarlshof show that the first Viking settlements were peaceful, which suggests that the islands were then sparsely settled. Unlike the circles and wheels that shaped Pictish art and architecture, the newcomers preferred straight lines and right angles. Their runic writing was made up of angular strokes, and their dwellings were square-ended. Grain, butter, wool and dried fish were stored in box-like outbuildings, while families lived in long single-story houses with people at one end and cattle at the other acting as a form of central heating from the far side of the living-room wall. In the living space, people slept on raised platforms on either side of the fireplace. Later these were sometimes enclosed to become bedrooms, or more often the beds themselves were given wooden walls (hence their name "box-beds"), which could be put up for privacy or removed for space.

Within a few generations, the Picts were barely a memory, and every bay, inlet and headland, every animal and sea-creature, every mood and belief, had received a Norse name. The bay was a *wick,* the inlet a *voe,* the headland a *ness. Dratsies* or otters hunted *sillocks* or young coalfish. A sullen man was *stut-*

set, a peevish man *snüsket,* and a weary man *drulset,* and in the shadows of the night, *trows, Brownies* and *njuggles* haunted the imagination rather than trolls, elves or water-horses. Their language was called Norn, and remained in use until the eighteenth century, long after Shetland and its southern neighbor, Orkney, had passed out of Norse control.

It was in 1468, that a Norse king, unable to raise a large enough dowry for his daughter who was to marry the King of Scotland, offered first Orkney and, one year later, Shetland in place of the money he had promised. The pledge was accepted, and since it has never been redeemed, both groups of islands passed into the kingdom of Scotland, and finally into the United Kingdom.

At this point, I should declare a personal interest. Whenever I introduce myself to a Shetland farmer, it is a good bet that he will comment, "Linklater? that's an Orkney name," and the remark will hang in the air, for an unmistakable rivalry exists between the two island communities. A Shetlander is said to be a fisherman who farms, while an Orcadian is a farmer who fishes, but the distinction goes deeper than it might seem.

From the sea, the Orkney Islands appear like a school of smooth green whales rising out of the waves. The hills slope gently and offer rich, easily cultivated land with good pasture for cattle and sheep. Sailing north, the first view of Shetland is of the huge, brooding cape of Sumburgh Head, and in a winter gale it wears a collar of white surf and above that a gray scarf of flying spray. Down its western flank, wind and tide boil up into a cauldron of racing water called "The Roost," which in 1993 trapped the oil tanker *Braer* and swept her onto the rocks, spilling her cargo of 85,000 tons of crude oil into the sea. Further off, the gigantic 900-foot-high cliffs of Fitful Head glower out over the Atlantic so menacingly that they inspired Sir Walter Scott to create a memorably grotesque witch to live there—Norna, otherwise known as "Mother doubtful, mother dread, dweller on the Fitful Head."

Where the geology of Orkney is easily worked sandstone, Shetland is mostly made up of deeply fissured gneiss, a granite-hard rock, which is often covered by a thick bed of peat; only in Dunrossness in the southern Mainland and on two islands, Bressay and Fetlar, is the soil naturally fertile. Long voes poke so deeply into the fissures that nowhere in its 90-mile length is there any part of Mainland more than five miles from the sea. Only about 45 inches of rain falls annually, but driven by the wind it can come flying in as hard as hail and

Laden with peats, Shetland ponies know the way home without reins or guidance and lead their masters, Lowrie Thomason and son Thomas. At low tide the ponies take the shortcut along the shore across the Sands of Tresta on Fetlar, circa 1930.

flatten a field of oats within minutes. Although winter brings snow, temperatures do not average much below 42 degrees, and the mean for summer is only a raw 57 degrees. No farm lies far enough from the coast to escape the thudding impact of an ocean storm or the corrosive effect of flying salt spray. Consequently, it is a rare Shetlander who succeeds in concealing his conviction that Orcadians lead a pretty soft life.

At least from the time of the sagas, the sea's close proximity has made it as important as the land in the islands' history. Indeed comparing "the quick silver harvest and the slow golden harvest," as the Orcadian poet George Mackay Brown expressed it, fish were often of greater value to Shetlanders than grain. In boats ranging in size from the sixern down to cockleshells only big enough for two oarsmen, they fished with long hand-lines for cod, ling and tusk in the rich but dangerous waters around the coast.

In one of the few saga stories concerned with ordinary, peaceful life, the charming and heroic Earl Rognvald ("Ronald"), who ruled the islands under the overall power of the king of Norway, took part in one of these fishing expeditions. Sailing back from Norway in 1148, he had been driven ashore in Shetland by an easterly gale. Some while later he encountered an old farmer on the beach near Sumburgh who was waiting to go fishing. When the farmer's companion who was to row the boat failed to turn up, Rognvald offered to take his place. As a poor man with a large family, the farmer depended on the sea to feed his household, and he accepted the offer eagerly enough but, to his alarm, Rognvald promptly took them out into the savage whirlpools and breakers of the Roost. "Just my luck to take you rowing with me today," the old man complained miserably, not recognizing his oarsman, "now I'll be drowned and my children will be penniless."

Naturally, having a hero at the oars, they survived, and the old man hauled up so many fish that when they got back to land and divided up the catch there was enough in Rognvald's share alone to feed the crowd of poor people who had come down to the shore. The saga's purpose was to show how an earl who was shortly to become a saint behaved, but to later, less romantic generations, the story is equally revealing as an illustration of the importance of fishing to a farming community.

That pattern of fish-supported agriculture continued with extraordinarily little change almost to the present day, but for a long time outsiders rather than Shetlanders made the best use of the sea's riches. European fishing was dominated by the

Dutch who had developed big, sea-going boats called *busses,* weighing up to 80 tons with round bows and sterns not unlike modern barges. Waiting for the fishing season to begin on the 24th of June, they would anchor in the hundreds—in the 1550s as many as 2,000 were reported—on the east side of the Mainland in the bay of Ler Wick, where a superb natural harbor lay protected from the open sea by the island of Bressay. The size of the fleet attracted traders from Hamburg and Bremen, eager to barter cloth and fishing gear in return for salt cod and herring.

Their presence also drew Shetland farmers anxious to sell fresh food to the foreigners. Trading stalls were set up on the shore of Ler Wick, and over the years a permanent settlement grew up which took the name Lerwick. It flourished, despite or perhaps because of warnings about "the great abomination and wickedness committed yearly by the Hollanders and country people repairing to shame at the houses of Lerwick." Brewers did a brisk trade in beer and, since the wool of Shetland sheep was prized for its fineness even then, farmers' wives came to sell "socks and other necessaries." If the local sheriff was to be believed, they were also guilty of "manifold adultery and fornication" with the foreigners. But condemnation did nothing to rein in the popularity of the place, and by the end of the seventeenth century Lerwick had grown to be the islands' largest town and its capital. In 1665, the seal was set on its position by the construction of a fort, Fort Charlotte, to protect it from Dutch men-of-war.

Until the eighteenth century, Shetlanders confined their own fishing to inshore waters where they caught saithe (a kind of small pollock) and shellfish. Then government policy made it profitable for home fishermen to venture further out to the *haaf* or the open sea where the herring were to be found. Now they needed a larger craft, and this was when the beautiful sixern first made its appearance. Since no trees grew on the islands, the boats or the precious timber to make them had to be imported from Norway. Thereafter from May until July, the fisher-farmers spent five or six days a week running their long lines from open boats up to 50 miles offshore.

It was always a risk to fish so far from safety, plunging up and down in waves which half-hid the faint outline of land on the horizon, and more dangerously the onset of a sudden squall. The fishermen had oars and a square sail, which was raised only when the wind was abeam or astern; if they were caught out by an unexpected storm, their one chance was to run before the wind. One sudden storm in 1832 drowned 105 men in a day, another in 1881 killed more than 60, and in every year the sea exacted its toll as surely as the bonxies.

The pursuit of the quicksilver harvest in such hazardous waters breeds a special sort of mental toughness and confidence in one's own judgement, which sometimes seems characteristic of Shetlanders as a whole. It may be unwise to generalize about 23,000 people when some get a quiet living in a commercial office in Lerwick, while others are out on the hill tending lambs in the teeth of driving sleet and snow. Yet to the outsider, newly arrived by sea or air from the south, the voices with their swinging Scandinavian cadences, and tendency to address any male of whatever age as "Boy!" carry the ring of an easy, unshakable independence.

In fact Shetland is the home of a quite particular, almost willful individuality. It is, for example, more than 250 years since the modern Gregorian calendar was adopted in the rest of Britain and 12 days were deleted so that dates might be realigned more closely with the seasons. Nevertheless, on the island of Foula, Christmas is still celebrated 12 days late, on the 6th of January. Why? For no better reason than not wanting to be told what to do by outsiders.

When oil was discovered in the North Sea in the 1970s, the oil giants, including Shell, B.P. and Amoco, selected the natural harbor of Sullom Voe, a deep fjord protected from the elements, as the place where the crude would be piped ashore. Despite their combined multi-billion-dollar influence and the unrelenting pressure of the government in London, the tiny Shetland Islands Council refused to let the project proceed until it had cast-iron guarantees on the protection of the voe's fragile ecology, and the assurance of a handsome royalty on every barrel of crude that passed through the terminal. Only those who have come up against a multinational company, let alone three of them supported by a national government, can appreciate the pressure that was brought to bear on Shetland, but the Council refused to back off. When I interviewed a senior executive at Shell almost a decade later, he still felt raw about the way they behaved. "It was the hardest bargaining I ever met," he said. "The terms they asked for were outrageous, but we had no place to go. The terminal had to be there, and they knew it."

It is a rare experience to feel sorry for an oil company. But today Shetland's school children, the elderly and sick, all benefit from the royalties paid at Sullom Voe.

The Sinclair family and friends relax during the harvest at Boats House, Whalsay, in the 1920s. Except for Sandy Jardine (holding Spark) and Jessie Rattar (white blouse), all are Sinclairs: in the back row from left: Kirsie, Robbie and Davey; Laura in her cap; seated: Eenie, Kirsie Jr., Sandy, Ellie and Willie.

In a sheepfold at the Lee near North Roe, Clara Tulloch holds a moorit (or brown) lamb. Here in the 1920s, sheep of different colors are still valued for their wool which is sorted by color, carded and spun without dyeing to be knitted into garments of rare hues. Later mills would favor wool that can be dyed many colors; so it happens that market demands encourage the breeding of white sheep whose wool can be colored any way at all.

Yet perhaps the most obvious example of the islands' temper is the one that lights up the dark winter sky on the last Tuesday in January each year. On a day when sunset comes barely seven hours after dawn, the festival of Up-Helly-Aa explodes with fire and singing, fancy dress and public satire, dancing and torchlit parades. Originally there might have been a pagan fire festival at this season to raise spirits against the gloom of a seventeen-hour night. Later it became part of the Twelfth Night celebration at the end of Christmas, but the event only took on its present form in the late nineteenth century when the blind poet, Haldane Burgess, wrote the *Up-Helly-Aa Song* (literally "the end of the holidays song"). Today its opening lines serve as the stage instructions for an annual display of atavistic pride in the islands' past.

> From grand old Viking centuries Up-Helly-Aa has come,
> Then light the torch and form the march and sound the
>   rolling drum.

The centerpiece of the procession through the streets of Lerwick is a copy of a Viking longship, accompanied by squads of guizers, or people in disguise. Escorting the ship, which is always described as a galley, are the Guizer Jarl and his squad wearing winged helmets and glittering armor, and carrying axes and shields. Behind them march thirty or forty other squads, made up of a dozen or twenty people dressed up as anything from extraterrestrials to tax inspectors—anyone, in short, who deserves to be celebrated or lampooned. Singing Burgess's song and carrying flaming torches, they march through the packed streets to the sound of bands playing and the scream and crash of rockets soaring up from the ships in the harbor. At midnight, the galley is hauled into an open sports field and there, on the Guizer Jarl's command, the torches are tossed into the longship, setting it alight. However modern the rite, it is impossible to witness the roaring flames leap from the belly of the dragon-headed vessel without a shiver of something like horrified exhilaration. Before the embers have died away, the guizers fan out through the town to spend the rest of the night performing and partying until the slow dawn appears.

Knowing of that determination to go their own way helped explain the small burlap bag hanging from a beam in the Isbisters' barn. It held the seed for next year's crop of *bere* ("bare") barley. This was the source of the golden harvest that the fisher-farmers used to drink to: "Health to man, death to fish, and good growth in the ground." For centuries bere barley and oats kept Shetlanders alive through the winter when there was no fishing, but today only the most stubborn would persist with cereal strains so slow-growing and low-yielding compared to modern varieties. For the Isbisters, however, it was as important to conserve it as to keep the Shetland breeds of livestock alive. "It's part of our inheritance," Mary insisted. "And if we lose it, we lose an essential part of Shetland's history."

Sometime in the autumn of about 1500 B.C., a farmer at Gruting in the Mainland was drying his barley when he allowed the peat-fire in the kiln to flare up too quickly so that ash covered the grain and spoiled it. The accident meant that 3,500 years later those twenty-eight pounds of barley could be dated and identified as being almost identical to the grains in the Isbisters' barn. And so close do past and present lie together in Shetland, the very working practices associated with the bere remain in living memory. In a neat little bungalow on the west coast of Mainland, looking across the sea to the island of Papa Stour, Dennis and Stella Shepherd even have them on film.

Almost a quarter of a century ago, the Shepherds were the resident church minister and schoolteacher on Papa Stour and, as a result of a declining population, the last full-time holders of the positions. No one could pretend it was an easy life. Apart from the wind and weather—"the islanders used to say the year consisted of three months of winter followed by nine months of bad weather," Stella laughed—there were the particular problems of living on an island three miles long by two miles broad, perched on spectacular cliffs, and lacking a proper harbor. Cattle had to be ferried to the mainland on a small boat and anything that could not be produced on the island was brought back the same way then hauled up a steep slope to the grassy fields where the fifty or so inhabitants lived. Electrical power did not arrive until the 1970s, and Stella recalled the excitement of one little boy when it first came through. "We're going to have an electric kettle," he declared, "then an electric dish-washer, soon we'll have *everything* electric, we'll even have an electric chair."

At eighth grade, children had to go to school in Lerwick, and few of them could resist the lure of the wider world. The reason can be found in the film the Shepherds made of life on Papa Stour. It captures a picture of the way people used to live all over Shetland. The fields were ploughed by Shetland ponies drawing a single-shared plough, then the earth was broken up by a man or woman pulling a harrow with a harness round their breast. They were followed by a sower scattering broadcast the seed for bere barley or oats. When the crop was ready for harvest, it was cut with a scythe or even a sickle, and then the harvested grain was threshed with a flail to separate the kernels from the chaff.

There was one tractor, but since most of the cultivated land was divided into strips about five or ten yards wide—a practice known as "runrig"—it was easier to work individual strips by hand. Livestock was either tethered close to the cottage or allowed to roam on the common land or *scattald* further up the hillside. Whenever the land-work permitted, sea-work began, taking boats out, setting lobster creels, repairing nets.

Curiously, despite the unremitting physical labor, the overwhelming emotion of the Shepherds' film—and Stella's evocative memoir of island life, *Like a Mantle the Sea*—is one of contentment. Lacking any radio, let alone television, the islanders made their own entertainment, gathering in each other's cottages to talk and knit socks and shawls. Major tasks were also communal, whether hauling boats above the high-tide mark or bringing sheep down from the scattald. Shetland sheep begin to shed their fleece in early summer and the crofters, rather than shearing them, pluck off the wool, a task called *rooing*. In its self-reliance and what might be called its sea-skills—reading the weather, boat-handling, and fishing—Papa Stour was a microcosm of Shetland itself. Dennis Shepherd's judgement on their time on the island was unequivocal. "Quite simply," he said, "those were the happiest years of our life."

Even on Papa Stour that way of life no longer survives today, but it was recognizably the traditional farm life described by the earliest historians of Shetland. Neighboring farms would be grouped round the most fertile land which was divided into "rigs" and owned by individual families. Further out, where the rough ground was suitable only for grazing, the land was owned by the whole community. Sheep and ponies were kept here, and a good wall separated it from the rigs. Even now, driving along the narrow one-lane blacktop roads that wind through the hilly landscape, the division can be seen clearly—a clustering of houses round neat green fields then, further up the slope, a sharp change to heather and rock dotted with sheep.

Bringing in the sheaves: At harvest time in the early 1930s, Frank Scott and Ertie Irvine hustle to beat the rain at Brae. Irvine's patient pony, beast of many tasks, stands ready in the traces. Next day he might bring peats down from the hill or carry his master all the way to Lerwick.

According to Andy Abernethy, who farms more than 200 acres of pasture and hill in the center of the Mainland, today's Shetland sheep, though healthier and better conformed, would not be so different from their ancestors. "There was never enough feed grown in Shetland to develop a big animal," he pointed out, "so I don't think they would have changed that much from the original sheep brought over by the Norsemen. You'll find sheep with a similar appearance in other places they settled, in Iceland and the Faroes, for instance."

Constant divisions to ensure that younger sons had part of a family farm meant that the rigs often became widely separated, and a holding of five acres might be scattered in five different locations. It was not an efficient way to work the land, and although seaweed was used as natural fertilizer, it was impossible to make any long-term improvements to the soil without reorganizing the runrig system.

The use of common land also had its drawbacks. With sheep belonging to several different families all on the same hillside, it was important that the right animals went to the real owners. Numerous laws were passed to encourage the use of identification marks and ensure that lambing, rooing and slaughtering took place with several people or preferably the entire community present. Nevertheless, sheep-rustling was always a problem. In the words of the Ordinance of Soid Brevet promulgated in 1299 by King Haakon of Norway: "Each taketh with dogs or otherwise what he can get that is not marked," it declared, "whether it be lamb or old sheep and whether it belong to him or no."

And it was no coincidence that this, one of the earliest pieces of legislation relating to Shetland sheep, should also have been concerned with Shetland sheepdogs, for there was no other way of catching animals that ran wild on the hill. These dogs were trained to hold, or *hadd,* sheep, hence their name "hadd dogs," and John Shirreff, who was evidently impressed, reported that "a properly bred dog always seizes the sheep by the off [right] foot, by the nose or by the ear." According to a later writer, Samuel Hibbert in 1822, it was a recognizable breed, and quite distinct from a mongrel or "running dog" which had no herding instincts. Unfortunately he gave no description of hadd dogs' appearance, though they must surely have been very different from today's delicately made Shetland sheepdog. Unscrupulous owners would use them to poach other people's sheep. As a result, the Soid Brevet limited the number of such dogs that any one person could keep, and imposed fierce fines for savaging sheep. Other laws went on to forbid a dog owner from going through the scattald at night, but clearly the problem of sheep-poaching never went away, because eventually an attempt was made to ban all dogs which did not belong to sheep-owners.

Probably the best policing came not from the law but the community who saw and knew everything that happened on the land. Each group of farms was called a township, and with six or seven people in every house, their populations varied from a couple of dozen to more than a hundred. Although one building continued to house people, grain, peat and cattle all in a row, the Vikings' more sensitive descendants separated the living quarters from the smell of the cattle-byre by putting a barn for storing grain and winter feed between the two. Nevertheless, some animals were still kept in the house. *Caddy* lambs, those that had lost their mothers and needed to be hand-fed, were kept by the fire, and in winter a young pig, or *grice,* was often tethered in the kitchen where it acted as a mobile refuse disposal unit. Close by, a tiny millstone turned by a stream ground barley and oats to meal, and any stray grain would have been gobbled up by the same feisty little black ducks that scavenged outside the Isbisters' crofthouse.

If the living was rough, so was the building, constructed as it was of stone and mortar, and covered with thatch or turf weighted down with straw ropes attached to stones. Chimneys were uncommon until the nineteenth century. Holes in the roof covered by a flap served well enough and window openings were filled with stretched lambskin.

Nevertheless, the Shetland farmers enjoyed one privilege denied to all but the nobility elsewhere in Europe. Until 1469 when Shetland passed into the kingdom of Scotland, the land in the townships was held under *udal* law, which governed Scandinavian property rights. This recognized the right of the person who worked the land to own it outright, free of any dues to a landlord. In an emergency when there was a fleet to be provisioned or a pilgrimage to be made, the ruling earls were not above overturning the law and then selling the right to ownership back to the farmers, but generally the principle of udal law was unchallenged. In an era when most people were bound to give feudal service to a lord or pay rent and dues for their land, the Shetlanders were as nearly free as made no difference, which helps to explain the feeling behind Up-Helly-Aa. History as well as geography draws Shetland toward Norway.

The union with Scotland was disastrous for the northern

Gathering the sheep on the island of Unst is a communal task by tradition. Despite the toil, a pause for a shared meal seems a picnic in a sheltered nook out of Shetland's almost constant wind.

isles. A succession of royal servants ruled Shetland and Orkney on behalf of the king until 1564 when the young Mary Queen of Scots gave them to her half-brother, Earl Robert Stewart. Both he and his son Earl Patrick, or "Black Pate," were gangsters who seized land and money without respect for any law, Norse or Scots. Although the greedier of the two, and the more hated for his habit of forcing people to work in the quarries, Black Pate at least left in Scalloway a handsome castle whose ruins still dominate the town. In 1615, he was executed for treason, but by then his Scots followers had been given large parts of Shetland, and other Scots followed, often as ministers for the new Presbyterian Church. With them came written laws which had the effect of confirming ownership in the hands of landlords, and making the farming-fishers their tenants.

Over the next two centuries, the behavior of these Shetland "lairds" would often bear an uncomfortable resemblance to a bonxie attack. As though it were not enough to charge a rent for the land their tenants lived on, the lairds even exacted a toll on the fish they took from the sea. Each boat had to land its catch at a place where the laird could count them and take his share, and on the tallest promontories, lookout points were built so that the laird's agents could see that no one attempted to evade the check.

The sea offered the only way out. In times of war, the Royal Navy came north to enlist men, and if they would not volunteer, the press gangs kidnapped them for service. In 1805, when Admiral Nelson went into battle at Trafalgar, it is estimated that as many as 3,000 Shetlanders were serving in His Majesty's ships. And in peacetime, whaling companies were equally keen to employ them, taking up to 2,000 men from the islands to the Antarctic every summer until the 1870s.

It was not impossible to prosper, and one who did was Arthur Anderson. Born in 1792, his first job was as a beachboy, turning the freshly landed cod which were laid out to dry on the rocks. He enlisted in the Navy, then worked for a shipowner before starting his own shipping business called the Peninsular Steam Company. In 1840, the company won a government contract for carrying mails to India and China, and was renamed the Peninsular and Orient. When Anderson died at the age of 77, he was not only the wealthiest Shetlander of his day but, much more enviable, the best loved for his earthy humor and generous purse. Today his company, known simply as P&O, is one of the world's largest freight-moving businesses.

During the eighteenth century, some of the richer families like the Giffords, the Bruces and the Hays gradually transformed themselves from landlords to traders. They built herring-curing stations, and they invested in rams to breed bigger sheep. Some found a market for ponies in the coal mines in the south, where they were used to haul coal in the pit-shafts far underground. Other lairds followed their example, but the investment was expensive. All the salt for curing, lines for fishing, boats for going to sea, not to mention the new rams and pony stud farms, had to be paid for in cash. Until then, most rents on the islands were paid in oatmeal or butter or fish or cloth, but if the lairds were to get a return on their outlays, the tenants had to become part of the cash economy.

The transformation enabled the lairds to establish sweeping power over their tenants. They lent them money to pay the rent and if necessary built new houses since the population was growing rapidly. They lent them money to buy boats, and if necessary provided fishing equipment for them since the market for herring was strong. Such loans were made at high interest. As security, the lairds insisted that any catch had to be sold to them, and then more often than not they used their monopoly to buy the fish at knockdown prices. The truck system, as it was known, was not unique to Shetland. In every company town where the same corporation owns the houses and shops and employs the people, a similar pattern can be found of high-priced goods, elastic credit and extortionate interest rates. The only difference was that being on an island, it was impossible to escape the system.

In the course of the nineteenth century, the lairds came to control the hugely profitable herring fishing. Some of the money earned from it and other enterprises went to build the substantial houses which made Lerwick one of the handsomest fishing towns in Britain. Some also helped bring on the darkest moment in Shetland's social history, for it allowed the lairds to invest in the one commodity that seemed to promise a decent return on the land. As the agricultural system had hardly changed since the Vikings, it was time for improvement.

From as early as the fifteenth century the one item that the Hollanders who came to Lerwick for the fishing always wanted to buy was Shetland knitwear, whether stockings or hats or rugs. Shetland women were famous for the speed of their knitting, and Shetland wool was noted for its fineness and springiness. Over the years, the demand for the islands' knitted goods increased, and in 1767 the list of exports included 50,000 regular stockings at six pennies a pair, and rugs and fine stockings at up to 30 shillings a pair, altogether worth £1,625 (when the average man's yearly wage was £40) perhaps $350,000 at today's values. In the nineteenth century the trade received a royal boost when Queen Victoria took to wearing the lacy shawls made by the best Shetland knitters. Throughout Europe, they became essential fashion wear for respectable ladies of a certain age. To the acquisitive eyes of the lairds, this market offered an opportunity too good to miss.

A small-boned Shetland sheep gave barely as much as two pounds of wool, but selective breeding had enabled stockmen in the south to produce heavier breeds of sheep, with names like Cheviot, and Leicester, and blackface, which carried more meat and produced up to three times as much wool as the natives. By crossing them with the Shetlanders, it was hoped to get the best of both breeds, fine wool and more of it. There was one drawback. The new sheep needed more pasture than the natives. The lairds' solution was drastic—the scattald must be fenced off and, in places, the tenants cleared from their fields. Thus was set in motion "the Clearances," the notorious pattern of legal eviction that helped to transform traditional Shetland society as drastically as it did life in the Highlands of Scotland. Using their power in a way that the islands had never known before, the lairds raised rents to intolerable levels, cancelled life-long leases and drove off families from farms they had worked for generations.

Where the clearances took place, the old community was effectively destroyed. In the island of Fetlar, half the land was cleared. The fertile lowland of the island of Bressay opposite Lerwick was cleared. The land round the old Norse center of government at Tingwall was cleared by the Bruce family, and upwards of 1,200 Cheviot and blackface sheep ran where townships had existed for a thousand years. In Unst the Edmonstons cleared families from 1,500 acres. In 1851 the population of the islands had grown to over 31,000; within a generation it had dropped by 10 percent. For those families who lost their farms and homes and were forced to emigrate, the clearances were a catastrophe, but the effect went much further, and to some extent persists today.

It should be said that there was something to justify the lairds' claim to be improvers. Shetland's agricultural system was

Well into the 20th century Shetlanders followed older ways: Their boats, as here at a quay in Lerwick, were driven by wind, their freight, the islands' docile beasts of burden. The ponies were as common and widespread then as pickup trucks anywhere today.

desperately backward and in need of change. The runrig system was inefficient; the common grazing on the scattald made it impossible to prevent the spread of diseases like sheep scab; and without greater fertility the land could not support so large a population. Indeed, it might be argued that people would have left the land even without the clearances. In 1886 the Crofting Act was passed, guaranteeing the tenants security against any further eviction, yet the rate of emigration accelerated faster than ever, so that in the next 80 years the population was reduced by almost half to a little over 17,000.

The reason that the clearances left so deep a wound was that they involved a profound betrayal of trust. Until then, the lairds and their tenants had been tied together for good and ill, relying totally on each other to make a living on the edge of the world. When the lairds turned on the tenants and evicted them from their townships against their will, it destroyed a bond which ran through life itself.

As they struggled to stem the hemorrhage of people and jobs from the islands during the first half of the twentieth century, Shetlanders looked increasingly for inspiration to another of the former Viking colonies, the Faroes. Formerly ruled by Denmark, the Faroe islanders had regained a sense of cultural identity by making Faroese rather than Danish the islands' official language. With that identity had come a sense of purpose which in turn fueled an economic boom.

For Shetland, re-creating the language was no longer an option. Norn had ceased to be spoken in everyday speech in the eighteenth century. Nevertheless, there was much in their cultural inheritance which remained to be salvaged from the effects of the clearances and depopulation.

In 1893, Jakob Jakobsen, a Faroese philologist, traveled to Shetland and began to study the remnants of Norn, eventually collecting a vocabulary of over 10,000 words. Many are still in current use today, particularly in farming and fishing. The categories of Shetland wool, for example, at a recent competition held in the village of Tingwall, would have been almost incomprehensible to an outsider. There were *moorit* or caramel and peat-brown fleeces; *catmogit,* a piebald chocolate and cream wool; and *shaela,* a silvery gray; had prices not been so poor, there might have been *bleagit, burrit, fleckit,* and *sholmit* fleeces as well, each one a precisely defined color combination.

Enough words remain to constitute a distinct dialect, and the pronunciation of many of them can fool a visitor into thinking he has landed in some part of Scandinavia. The word for a sheepfold is *crø* which sounds like "crurr," while a fisherman's basket is a *budie,* pronounced as if it were spelled "beeudy." A distinctive way of saying "th" as "d" means that "the" sounds like "da," and words like "father" and "mother" come out as "fedder" and "midder," while the familiar form "thou," which is also still used, becomes "doo." In fact, listening to a conversation between two Shetlanders in a Lerwick pub or shop, it can be hard to know that they are speaking English at all, and most islanders learn to make a conscious effort to modify their speech to make it comprehensible to visitors.

Inspired by Haldane Burgess, many Shetland poets consciously used this speech which they had all spoken as children. A cautionary verse by one of the best of them, Rhoda Bulter, about the irresponsibility of little Jimmy, *"peerie Jeemie,"* gives an idea of the richness of the dialect.

Da coarn's been truckit be twa unken kye,
Da sheep is come oot o da park at da banks,
An da aald moorit almark is broken her branks.
Whaar, tell me whaar, can be yun boy o wirs?
He's lyin soond asleep in a swerd o green girse.

(The corn's been trampled by a pair of strange cattle,
the sheep have got out of the field on the hill,
the old brown ewe that always strays has broken her halter.
Where, tell me where, can be that boy of ours?
He's lying sound asleep on a sward of green grass.)

Yet until recently it was a minority interest. In school, children were still being taught to cut out dialect words altogether, and when local radio began 20 years ago, the first announcers were chosen for their ability to speak BBC English rather than the speech their audience used. Much the same was true of traditional music. In the old townships the fiddle was the instrument played for dances and weddings, but with the emigration of so many young people, there was a risk that the old tunes would be lost. By the 1960s, many of the intangible influences that make up a community's culture seemed to be fading.

In farming too, the distinctive breeds of sheep and cattle, ducks and geese, were in danger of being entirely superseded by the varieties first introduced by the improving lairds. The old runrig system virtually came to an end in the nineteenth century, and many small crofts left unoccupied by the falling population were amalgamated into larger farms. With fertil-

On the shingle beach where their ilk are as nearly at-home as on any grassy hill on the Isle of Noss, young lambs nuzzle Jeemie Jamieson. Shetland sheep know the rhythm of the tides on these islands and often forage for seaweed along the shore.

izers, drainage and high-protein feeds, any farmer could run Cheviots and blackface, Aberdeen-Angus and Holstein. There was no need for the tough little breeds that could stay out all winter and live off the thin hilltop pasture. Sometime in the first half of the twentieth century Shetland's pigs, robust and lively beasts that lived in the wild, disappeared. The small black cattle became so cross-bred they were scarcely recognized as a distinct variety and, although the sheep survived thanks to the establishment of the Shetland Flock Society in 1924, the breed was scarcely thought of as a commercial proposition.

In the 1960s the discovery of oil in the North Sea, followed by the choice of Sullom Voe for the main on-shore terminal, triggered an economic boom which was to transform the islands. In fact it would be easy to conclude that oil alone saved Shetland. The raw numbers certainly point that way. Between 1971 and 1981, the decline in the population not only ceased but was dramatically reversed. From little more than 17,000, it rose by no less than a third as jobs were created in the oil industry itself and in services from harbors and airports to taxis and shops. With it came renewed economic confidence, which helped in turn to foster a new pride in Shetland culture as the younger generation realized that the search for employment need not now take them away to job markets in the south.

However, it would be a mistake to ignore the influence of other, less tangible factors. The growing popularity of Up-Helly-Aa had signaled an increasing enthusiasm for the islands' past. And pride in the dialect was already reviving. Once confined to the playground at school, or the fields and streets, it began almost insidiously to penetrate business meetings and classrooms and, reflecting this progress, Shetland announcers started to make their voices heard on the radio station.

In music most of all, there had been a burst of enthusiasm for Shetland tunes and dances, triggered by The Forty Fiddlers, a group set up by Dr. Tom Anderson to play traditional fiddle music. One of his earliest students was Aly Bain, whose virtuoso interpretation of the islands' distinctive style of playing has brought him an international following. Even in fishing, the old mainstay of the economy, enterprises like the Whalsay fishing co-operative had begun to show the way forward by investing in larger vessels and better processing facilities.

The consequence was that when the offshore oil industry surged into Shetland, it encountered a society which, however drained by emigration, still kept a lively sense of who it was and where it came from. Above all, it retained to an extraordinary degree its old cohesion. To the incomers, adjusting to that close communal identity was often hard. One small example was the business of visiting a neighbor. For the newcomers, it seemed only polite to phone ahead to find out if it was convenient to call round. For the Shetlanders, this seemed like showing off, as though the visit was so important that time was needed to prepare for it. They were used to neighbors who dropped in unannounced and, because no one ever locked the door, usually without even knocking—behavior which struck the newcomers as rude and intrusive.

Enormous adjustments had to be made on all sides, but it is fair to say that a quarter of a century on, the oil industry looks much more like part of Shetland society than Shetland seems part of the oil industry. Perhaps that is not so surprising. What has formed the islands more than any other influence is their position, far to the north, and far from other land. Whoever lives there must live with the wildness of wind and weather, and their unchanging demand for self-reliance and resilience.

When Scotland was offered the chance in 1997 to vote on having its own parliament with tax-raising powers, Shetland defiantly went its own way and, unlike every other constituency in the country, voted to reject both parts of the proposition.

To those familiar with the islands' historical distrust of rule from Scotland and the profound sense of independence that geography breeds in its inhabitants, any other outcome would have been astonishing.

Increased prosperity has transformed the infrastucture of Shetland with new roads, frequent ferries, and daily flights linking islands and outlying areas. The earning power of fishing, tourism, knitwear, and ancillary services has grown so much that oil's slow decline as an employer has scarcely been felt. Seven of the largest fishing trawlers in the world are owned by seven families on the island of Whalsay, and in the latest statistics from the Shetland Islands Council, the output from catching, farming and processing fish accounted for almost a quarter of the islands' economy. Only farming has not shared in the general boom. In fact, agriculture recently suffered its worst depression in more than a generation, but paradoxically there could have been no more powerful incentive in reviving interest in Shetland breeds of livestock.

For Jim Johnson, whose farm on the beautiful island of Muckle Roe was once part of the Gifford lands, the logic of falling prices was inexorable. It made no sense for Shetland beef farmers to try to compete with southern producers by breeding the big Holsteins and Charollais which required so much artificial feeding. It was time to bring back the small, black Shetland cattle. "They are ideal for this part of the world," he pointed out. "They can stand up to the weather, and they'll do well on grass and hay which reduces feed costs. They develop more slowly than modern breeds, but the flavor of the beef is far superior."

On a November day when rain slanted in on the wind like bullets, he walked through his herd. The Charollais which he used to keep would have had to be inside. The Shetland cattle had gathered in the lee of a hill, but otherwise they seemed almost oblivious to the weather. "All the same," he admitted, "I worry about them in these conditions. They're such nice, gentle beasts, they don't deserve to be out when it's like this."

In the late 1990s wool and sheep prices plummeted to the point where the price of a lamb sent to market on the mainland of Scotland was less than the cost of transporting it there. In those circumstances, the virtues that Jim Johnson found in Shetland cattle were very like those that Andy Abernethy valued in Shetland sheep. Even though he could not sell them, his Suffolk crosses still had to be fed. The native sheep, on the other hand, had long been adapted to hill grazing. "It's the right breed for us today" he pointed out. "The Shetlanders can get by on what they find on the hill. They were bred for that. And the quality of their wool means that there's still a demand for it even at times like this."

At the Tingwall wool competition, the judges examined the wool by holding it up to the light to examine its fineness, and by tugging the strands to test their strength and springiness. The weather that burst upon the road as I left the Abernethys' farm demonstrated just why that quality was needed.

An extra gust from a buffeting wind out of the northwest lifted the dark fleece on some catmogit sheep grazing by the road. There was a momentary gleam of lighter wool below before the outer layer fell back, and the gust skittered onward, darkening the grass on the hill and continuing across the waters of Southsound Voe, combing them into a flurry of white horses. It was cold enough to make the eyes water. Then for a minute the low winter sun streamed through a gap in the clouds throwing a light, as white and blinding as a magnesium flare, and left the faintest sense of warmth on the face. Almost before it had faded, the clear outline of the hills beyond the Abernethys' farm was blurred by a purple and gray curtain which quickly blanked out the fields and arrived at the road in a hammer of stinging hailstones. Through all the changes, the sheep barely paused in their grazing, and the white hail lay unmelting on the dark fleece like the onset of old age. Whatever the weather thrown at them, the tightly crimped wool was proof against it.

There is no flock book to govern the breeding and character of Shetlanders, but their ability to cope with these same conditions is even more remarkable. Extremes of latitude teach lessons not to be learned elsewhere on the globe. At 60 degrees north, the elements peel away the deepest layers of comfort, the weight of winter darkness disturbs the calmest mind, and a visitor sometimes feels that mere survival is a triumph. Over 3,500 years, however, the fisher-farmers of these islands have adapted to their environment, and quite unconsciously forged for themselves a unique identity. Like the animals in their farmyards, and the birds that nest on the cliff tops, they have learned to live on the edge of the world. They have learned not only to thrive in these extremes, but to take in their stride the extra, unlooked-for stroke of misfortune, the bonxie attack, that would surely crush anyone less resilient.

A prime Shetland Sheepdog, circa 1900.

THESE ISLANDS' ANIMALS

A classic pair: Mare and foal in a classic pose.

Shetland Pony

A Most Appealing Breed

Valerie Russell

Shetland ponies are the smallest of the nine breeds of native ponies that have developed over thousands of years in the island, mountain and moorland areas of the British Isles. They measure at maturity no more than 42 inches tall at the highest point of the withers (i.e., the third to ninth vertebrae between the neck and the back). Uniquely among British native ponies, their height is traditionally given in inches, not hands. They may be any color except spotted, and are the only British native pony breed to include piebalds and skewbalds. The Shetlands' small size, together with their sturdy, short-legged build and profuse mane, tail, and forelock, make them arguably the most easily recognized of all equine breeds worldwide. They are famed for their great strength, versatility, surefootedness, and delightful if slightly mischievous temperament. Their physical features are classic examples of adaptation to a very harsh environment, and fully to appreciate this wonderfully hardy breed, it is helpful to understand the environment of the Shetland Islands.

In the group of more than 100 islands, the three largest—Mainland, Yell and Unst—offer a rugged landscape of comparatively low hills and rough moorland, interspersed with valleys which may contain pools or larger lochs (lakes) of brown, peat-stained water. On both hill and moorland there are rocky crags and outcrops, as well as soft, boggy areas, with vegetation consisting of a mixture of acid-loving heather, rushes, reeds, sedges, mosses, and rough grasses. Below the surface of all the islands are deep layers of peat.

Many of the valley bottoms have been reclaimed from the rough vegetation and reseeded, offering a more fertile environment for farming. The landscape is treeless, and although frequently described as bleak and windswept, at certain times of the year it presents a stunningly beautiful mosaic of the rich and varying colors of the vegetation, with all shades of greens, reds, and browns. No place is more than a few miles from the sea, and the rocky, spectacular coastline is indented with narrow inlets known as voes.

The position of the islands, some 200 miles west of Bergen in Norway, is deceptive in terms of climate. The north-flowing warmer sea currents of the Gulf Stream moderate the winters to an appreciable degree, so that instead of the extremely low temperatures and all-embracing snow of Scandinavia, average temperatures of Shetland winters are similar to the northern areas of mainland Britain, although the summers are cooler.

Nevertheless, winter conditions are extremely demanding on ponies and people alike, with frequent ferocious gales, blizzards, and heavy rain blowing in from the Atlantic, and with a severe wind-chill factor. In 1633 a visitor to the islands wrote of "the Great Winds of Winter. . . . I can speak by experience, being blown down flat to the ground by the violence of the Wind. I was forced to creep on my hands and knees to the next Wall, and going by the Wall, got into an house." A century and a half later, a geological surveyor recorded "During the first three days of January 1787 it blew a violent gale. . . so furious that the sea broke over headlands 120 feet high and at one place, a stone, computed at 12 tons weight, while laying 90 feet from the sea, was raised out of its place and thrown two fathoms farther from the shore." And a century and a half after that, during World War II when the future English historian David Howard was serving in the Royal Navy, "On 10th November an exceptionally violent storm broke over Shetland . . . . [and soon] reached hurricane force. It blew for five days and nights, often with a velocity of well over a hundred miles an hour." The weather in Shetland has not modified with the passing of time.

So, how are the ponies adapted to these demanding conditions? Their size is clearly important in a treeless landscape in which shelter is to be found behind rocky outcrops, in hollows, and since the coming of man, behind the long, low stone houses and dry-stone walls that separate the fields. The smaller the pony, the more shelter it is able to find. Similarly, the food available, never highly nutritious or abundant, is particularly scarce and of poor quality in winter. Thus the smaller ponies, requiring less food, have been the ones best able to survive.

Further, the typical outline of the Shetland, with its short legs and compact body (in which the length of the legs equals the depth of the body behind the front legs), and the small neat ears, offers the minimum surface area from which heat can be lost. This is in accordance with Allen's Law, which states that "body parts such as legs and ears tend to be shorter and smaller relative to overall body size in animals living in cooler climates."

Most mountain and moorland ponies, including Shetlands, have relatively smaller, shorter heads than horses. This adds to the delightfully alert, lively expression which contributes so much to their charm. More importantly, it contributes to their survival. The ponies have a markedly deep jaw in comparison with horses, and this allows for the insertion of the exceptionally strong, long-rooted teeth needed to deal with the coarse, fibrous herbage on which they live. Neither may the head be too small (a feature being bred into some native breeds to suit show ring fashion), as the nasal cavity must provide a sufficient area of highly vascular lining membrane to allow cold air to warm before reaching the lungs.

Finally, the coat is a model of adaptation to the environment, as demonstrated in the research on coat structure and other survival features done in the 1940s and 1950s by Professor James Speed, Lecturer in Comparative Veterinary Anatomy at the Royal (Dick) Veterinary College at Edinburgh University, and Miss M. G. Etherington (later Mrs. Speed). Much of their work was done with Exmoor ponies, but it applies equally to the other native breeds.

The Shetland's winter coat has two distinct layers. The undercoat consists of fine, springy hairs, measuring an inch to an inch-and-a-half in length, and projecting from the skin at a slight angle. Among them, especially on the flanks and back, are flakes of a greasy material which apparently helps to entrap air and thus adds to the coat's insulating efficiency.

Lying over the finer hairs, and growing in the same direction, is an outer coat of much stronger, wiry hairs, of sufficient length to cover the undercoat. The insulating properties are proved by the appreciable periods of time during which snow lies unmelted on the ponies' backs. The outer layer of the coat is cast in late spring, and grows again in autumn.

While native ponies can survive extremes of weather, they can do so only if their coats are waterproof. They cannot stand cold if it is accompanied by rain which penetrates to their skin. Thus waterproofing of the coat is absolutely vital, and in wet weather a survival feature becomes apparent as it prevents rain or melted snow from reaching the skin. When wet, the appearance of the winter coat alters, and it is apparent that the hair lies in definite streams *(flumina pilorum)*. Also, the whorls or vortices of hairs sited at various points on the body—such as the prominent one on the flank just in front of the thigh—become much more obvious. Groups of hairs in these streams aggregate at their tips, forming triangles, the apices of which point downward. Rain is led off the body surface by these apices, which thus prevent it from penetrating the coat and from reaching the less-protected parts of the anatomy, such as the underbelly and the region beneath the tail.

The most noticeable vortex, lying at the rear of the flank, is thought to be the most important. The hairs at the upper part of this, just forward of the point of the hip, fan outwards, and then turn either frontwards or backwards. These hairs join the stream from the upper part of the body, and lead water away from the inside of the thigh and the underside of the abdomen. Other vortices and whorls, such as the one sited between or slightly above the eyes, play similar but probably less important roles. Observation suggests that in the true native pony living under natural conditions, misplacement of these structures results in less hardiness.

The profusion of hair in the Shetland's mane, tail, and forelock also provides protection from the elements. In a pony living in its native habitat, the forelock may reach right down to the muzzle, giving protection to the eyes, while the mane can reach the ground. As horses and ponies turn their backs to the prevailing wind with their tails well tucked in during bad weather, the importance of plenty of hair in the tail is obvious. Shetland ponies also have what is known as a "snow chute" at the root of the tail. This is a thatch of short hairs (a continuation of the hair stream running the length of the back) which fans right and left from a vortex sited at the very end

Laden with peats, Lowrie Thomason's ponies lead the way down the hill road to Lambhoga, on Fetlar, circa 1930. True to form, the string of ponies, all mares, follow their leader when "flittin' the peats," i.e. carrying bricks of fuel cut from beneath the island sod.

of the spinal column, and which leads water away from the dock area.

Shetlands, like other native ponies, often have low-set tails, due to the last group of vertebrae being set at a downward angle towards the tail—a feature known as "goose-rumped." This too is an advantage, as it enables the tail to continue on the downward slope, and enables it to be tucked in very close in bad weather. It is, however, a feature much criticized when seen in the show ring; here is another example of the quirks of fashion being at odds with the demands of survival!

PONIES' ORIGINS AND OLD USES

So, the ponies are well adapted to their island home, but how long have they been there, and what are their origins?

No fossilized evidence of ponies exists in the islands' rocks, but bones have been found in different layers of excavations at Jarlshof near Sumburgh, on the southern-most tip of the island of Mainland. Jarlshof was occupied by Bronze Age Man from about the sixth to the first century B.C. Reports compiled by Miss Margery Platt, M.Sc., of the Royal Scottish Museum during the 1940s and 1950s described the remains as "being sparsely scattered throughout the levels. Bone measurements correspond favorably with those of the Shetland pony as it is in modern times, therefore it is possible that they belong to this type." Miss Platt described bones found in higher levels of the excavations as coming from Viking settlements of the early ninth century A.D. These were also pony bones, but of animals larger than present-day Shetlands. Finally, more bone fragments of a similar size to those of the Viking era were unearthed from late Norse middens of the eleventh to thirteenth centuries.

While the foregoing is positive proof that ponies have been in Shetland at least from the Bronze Age, their origin and ancestry are still very much a matter for speculation. The sequence and variations in size suggest that there may even have been multiple origins of small horses in Shetland. Current work on native ponies using new methods of protein analysis and DNA fingerprinting may yet explain their origins.

For the moment, however, the belief is that ponies arrived in Shetland before the end of the last Ice Age by crossing the ice fields or land bridges which then connected Britain to Europe. Professor Speed suggested that the earliest ponies could have been of two types—a "mountain" pony standing about 12.2 hands and a sturdier "cob" type of about 13.2 hands. Bones of both types have been found in mainland Europe and in southern Britain dating back some 10,000 to 12,000 years. The subsequent diminution to the size of the modern Shetland might be explained either by a biological law stating that animals developed in isolation become smaller as one stage of their evolution, or as the result of environmental adaptations.

Professor Speed further proposed that the "mountain" ponies came to Britain via the "Oriental" route from North America, via Asia Minor and the Caspian area to the Mediterranean, thence up through the Iberian peninsula. The larger "cob" type could have reached Britain by Siberia and what is now Scandinavia.

This theory is supported to some extent by Dr. Gus Cothran at the University of Kentucky College of Agriculture, who has studied genetic relationships between various breeds of ponies. His work suggests that Shetland ponies and Icelandic horses (which have certain characteristics in common) might have a shared ancestry, possibly with the extinct Celtic pony which originated around the Caspian Sea, and some type of Norwegian equine, similar perhaps to the Norwegian fjord horse.

Artistic evidence of the presence of small ponies in Shetland has emerged with two separate discoveries of carved stones dating from the ninth century A.D. The first, found in 1864 and called the Bressay Stone (as it was found on the island of that name), depicted a pony and rider. The second was a group of three stones found beside the old church at Papil on the island of Burra. On one stone was a carving of four unmounted figures and a hooded figure riding a pony. Calculations have indicated that the ponies in both instances stood about 43.3 inches—just a little bigger than the maximum permitted height of the modern Shetland pony.

The earliest known written reference to the ponies dates from 1568, when one Ubaldini wrote in *Description deal Redo di Scotia* that "Their horses are very small and tiny in stature, not bigger than asses, nevertheless they are very strong in endurance." This is the first of a number of writings commenting on the size, strength, and endurance of the ponies.

These qualities have been employed to good effect by the people of the islands for centuries. Historically, the majority of Shetlanders followed a form of subsistence agriculture known as "crofting." Under this system, each crofter occupied his croft house, which stood on a few acres (at most) of "in-bye" land.

Jeemie Sinclier and his horse Jess pause at Sullom, Northmavine, as Geordie and Betty Jamieson stand proudly by in the 1930s.

Different portions of these few acres were used for cultivating such crops as would grow in the acid soil, for grazing the house cow which provided milk and butter and, during the worst of the winter, for accommodating the oxen used for ploughing.

Each group of crofters also had grazing rights over large areas of hill and moorland, the *scattalds*, where they turned out their sheep and ponies. In a life of extreme poverty, the crofters looked after the ponies to the best of their ability, supplementing the poor grazing with potatoes and straw when they could. Often food was short for man and beast alike, and there is no doubt that some ponies did not survive the harsh winters. Those that had access to the seashore probably fared better, as they could (and as they do to this day) supplement their diet from the vast banks of seaweed flung ashore by winter gales.

The ponies played a vital role in the crofters' existence. They were the sole means of transport, and because there were no roads in the islands until late in the nineteenth century, "transport" meant riding. Incredible though it may seem, there are written descriptions of these ponies carrying the crofter and his wife. In 1701, John Brand, in *A Brief Description of Orkney, Zetland, Pightland Firth and Caithness,* wrote "Yea there are some whom a man can lift up in his arms, yet they will carry him and a woman behind him 8 miles forward and as many back." They were then, as they are now, noted for their sureness of foot; as a naval surgeon reported in 1780, "We were furnished with little horses and set out over hill and moor, rock and stone. . . . We trotted along brinks of dreadful precipices where I would not venture to trust myself on the best hunter in England. . . ."

The ponies' other main use was "flittin' the peats," i.e., carrying fuel. The peat banks from which the crofters had the traditional right to dig and remove peat were often some distance away on the hills. The peats, each measuring roughly 10 inches by 6 by 4, were dug with special tools and laid out to dry sufficiently to allow them to burn, but not too quickly. To carry them down to the houses, the ponies wore special tack, called a "bend" consisting of a wooden pack saddle ("klibber") with two sideways-pointing horns. (These were made from driftwood collected along the shore.) The peats were loaded into flexible baskets ("rivakeshies" or "kishies") made from woven rushes or grasses, and kept in place by nets ("meshies" or "mashies") which passed round the kishies and hooked over the saddle horns. Great care was taken to pack either moss or heather under the front of the klibber to stop it from rubbing the pony's back, and a well-padded rope led from the rear of the klibber and under the tail to prevent the bend from moving forward as the animal walked downhill, and causing discomfort. As an old crofter who had worked with the ponies as a boy explained recently, "Everybody was extremely careful with their ponies. They had to work, and they were your strength, so you looked after them very, very carefully."

Only mares were used for flittin'. They were rounded up on the scattalds and tethered "in-bye" close to the houses for the duration of the flittin' season. Mares with foals were tethered by the foot just above the fetlock so the foal did not become entangled in the rope. Flittin' in these northern islands where the summer nights are never really dark sometimes started as early as two o'clock in the morning, so that the ponies would not have to work in the hottest time of the day. The work usually involved the whole family, including the children, whose task was to lead the strings of ponies up the well-worn paths to the peat banks, sometimes an hour's walk from the crofts. From time to time, as new banks were opened, the ponies would establish a new path, walking carefully with their noses to the ground, avoiding soft and boggy areas. So adaptable were the ponies that they had only to be led to a new bank once; after that they would go to it without further direction, as I was told by two elderly former "pony boys" who worked the peat banks, Messrs. Tony Priest and Willie Spence. Once at the bank, the peats were loaded, and the children escorted the ponies back down to the crofts, where the peats were stored in carefully constructed stacks.

A string usually consisted of six ponies, with an established older mare as leader. It must have been a charming sight in good weather, as the foals, running loose, accompanied their dams, sometimes taking a rest amongst the heather, and rejoining the string as they made their way back down the hill. The number of journeys made each day was fixed, and the wily old mares knew it. When waiting to be loaded for the last trip down, instead of standing quietly or grazing patiently, they would become restless, and more often than not would try to set off down the hill before they were fully laden.

Although most of the peat ponies had accompanied their dams to and from the banks as foals, they still needed to be taught the job of flittin'. This was done when they were two or three years old. Once accustomed to the weight of the "bend" on their backs, they joined a string, and initially had a long rope attached to their halters in case they suddenly decided to make a break for freedom. In such an event, and with a bit of luck, the peat boy would be able to grab the rope as the young pony went galloping past!

Although mechanization came to the peat banks in the 1950s, it is still possible to walk along many of the tracks trodden by generations of ponies long gone.

In addition to flittin' and still wearing their "bends," the ponies would also carry in hay and straw from the small fields; on these occasions they had a special addition to the head-collar which prevented them turning their heads to eat their loads. A few Shetlands were used for ploughing, harrowing, and, once roads were built, for pulling carts, but these tasks were done chiefly by oxen or by cross-bred "work ponies."

As crofters were also fishermen, the ponies made their contribution to this way of life as well. Hairs from their manes and tails were woven into fishing lines—a practice which continued well into the twentieth century. So valuable were these lines that, in an effort to stamp out theft, officials sometimes checked that every crofter could account for the origin of all horsehair lines in his possession.

TO THE MINES OF ENGLAND

Thus for centuries, Shetland ponies lived an almost entirely rural, agricultural existence as riding and pack ponies. Then in 1842 an Act of Parliament passed in faraway London dramatically changed the lives of many ponies and, in the long term, the future of the Shetland breed. The Mines Act prohibited the employment of women and children under the age of ten in coal mines, where they had been engaged in moving coal from the coal face, initially by means of baskets on their backs, then on sledges, and eventually, immediately before the passage of the Mines Act, in wheeled carts.

Deprived of this traditional labor force, the mine owners cast around for an alternative—and found it in ponies. From that time, until as recently as the post-World War II era, hundreds of thousands of native ponies worked in the mines—

the smaller ones such as Shetlands in the narrow, low-ceilinged underground galleries, and the larger breeds on the surface. The mine owner who was to have the greatest influence on the breed was the fifth Marquis of Londonderry, who had extensive mines in County Durham in northeast England. He started using Shetlands about 1850, and established a stud in Durham with ponies imported from the islands as he became increasingly concerned at the appalling standard of the ponies being bred.

This decline had been taking place over an extended period for several reasons. Firstly, each little community of crofting families was more or less isolated from neighbors, as were the scattalds on which they turned out their ponies. As a result, inbreeding of the ponies was rife, and the standard inevitably declined. In addition, the crofters, when given the chance of selling a pony for export, inevitably sold off their best animals. Finally, the mine owners required not mares but colts and stallions to work in the pits—so the stock was further denuded of the best males that might have provided useful outcrosses. The earliest photographs of the ponies of the late nineteenth century show many miserable little creatures, with shallow bodies and long, spindly legs—so much so, that it is all the more remarkable they could carry an adult rider.

Although some attempts had been made by a few of the larger landowners in Shetland to improve the breed, they were not widespread. So, Lord Londonderry resolved to reverse the decline, and established studs on the islands of Noss and Bressay (just off Mainland near Lerwick) with the sole aim of breeding pit ponies. These efforts were overseen by his Durham stud manager, Robert Brydon, who was also a veterinary surgeon.

The aim was to produce ponies "with as much weight as possible and as near the ground as can be got." To achieve this, Brydon and his Shetland manager traveled throughout the islands and initially bought some 200 colts and stallions that approximated the above description as nearly as possible. Unsurprisingly, many were found unsuitable, but six stallions were finally selected as foundation sires for the stud. Two of these, Jack 16 and Prince of Thule 36, were, in due course, to have a profound influence on the entire Shetland breed.

To achieve their ends in the shortest possible time, Brydon and his men in Shetland practiced inbreeding of such closeness and on a scale that would horrify modern breed-

Robbie and Willie Sandison, astride their pack ponies, take a happy break from the toil of transporting peats on Unst about a century ago.

The land slopes away toward Natal Cottage in the fertile landscape at Easter Quarff, 1936. Grain ripens in the fields, a pony awaits its rider, a crofter walks a path, and all is right in this realm.

ers. But they soon produced a pony that became known as the "Londonderry type." This was described at the time as a pony with more bone and substance and improved limbs, but it was criticized for a tendency to straightness in the shoulder and lack of withers. However, the latter two features, while undesirable in ridden ponies, are normal in draft animals required to pull loads, which, up until this time, Shetlands had not done to any extent.

In addition, the improvement in bone and substance, while attributable in part to selective breeding, was undoubtedly aided by an improved management regime that the average crofter simply could not afford. Instead of foals staying with their dams throughout the winter, Londonderry foals were usually weaned in November, thus relieving the mares of the strain of producing milk for them during the hardest weather. The old method had often resulted in mares only foaling biannually. In addition, sheds were provided as shelter, and the mares were given supplementary food in the form of hay. The improved management resulted in many more mares foaling every year.

The young colts and fillies were turned out on the hills during the summer, still separated from the mares, who were kept near the shore, where they ate quantities of seaweed from which they gained important minerals for good growth of bone. Fillies were sent to selected stallions, usually at the age of three, and it was of significance to the breed as a whole that these stallions were made available to mares belonging to the local crofters—so an overall improvement of their ponies also took place.

The foundation of the Londonderry stud led to the founding of other studs in Shetland and in mainland Britain. Furthermore, the majority of influential studs set up during the late nineteenth and early twentieth centuries were founded on Londonderry bloodlines and, to this day, the majority of Shetland ponies trace back to the original six Londonderry foundation stallions.

The improvement in breeding and management practices set new standards, as Dr. Charles Douglas and his wife, Anne Douglas, wrote in the first book on the breed, *The Shetland Pony,* published in 1913: "The consequence was a degree of breed improvement which is perhaps without parallel as the result of less than thirty years of breeding and management." Had the previous decline of the ponies been allowed to continue, with the crofters selling off much of their best stock, the future of the breed must have been in doubt. That the decline was halted was due almost entirely to the Londonderry Stud.

However, the stud was founded with purely industrial objectives, and many thousands of ponies from the islands spent long years working underground. They were easily broken to harness, as Robert Brydon recorded, "The Shetlands' docile temperament enables them to be trained for pit work in almost as many days as it takes weeks with Welsh ponies." Their task was to pull wheeled tubs, which ran on rails, to and from the coal face. Their work rate was prodigious. It was estimated that a pony would walk over 3,000 miles underground in a year, and move an equal number of tons of coal.

Proof of the remarkable strength of Shetlands was shown by achievement tests conducted by a German researcher, Dr. Flade. He showed that a Shetland can pull some two-and-a-half to five times its own weight, and he explained that the pony, when pulling a load, pushes himself forward using his hind legs (which is normal for a riding animal), instead of first lowering the forehand (the head, neck and shoulder) as in draft horses that are bred to the purpose.

Conditions underground, with the low, narrow galleries, were hazardous for miners and ponies alike, and accidents were commonplace. In time, the ponies were equipped with specially designed headgear which protected the top of the head and the eyes. The cramped conditions made turning in the galleries difficult, but, resourceful and adaptable as ever, the ponies overcome this, as is described in John Bright's book *Pit Ponies:* "A pony would tuck his head between his legs, turn slowly round till his neck touched the sides, then bring his back legs in and spin like a top."

It was always believed that the ponies had a sixth sense which warned them of imminent danger, and there are many tales of them refusing to go forward toward a point where a roof fall would soon occur. The ponies were led by miners known as pony boys, and there are also stories of ponies saving their boys and vice versa. Bright describes one such incident told by a miner. "I was running down a steep hill to turn a railing in order to allow the full tubs to proceed, and as I bent down to turn the 'point,' I slipped on a wet sleeper and fell down with Fido and the tubs almost upon me. Obviously realizing that the tubs would hit me, Fido shoved his back end

At Westing on Unst, Maymie Smith checks the leaders of her train of Shetland ponies to pose, her smile as bright as the wild flowers at her feet. From spring to autumn, when Shetland men were fishing or at sea, the heavy work of farming, cutting and carrying peat, tending the flocks, and maintaining the crofts fell to women and children as a matter of course.

A handsome Shetland stallion squares up for his handler one last time before a livestock show around the turn of the last century.

into the breach and shot his front legs straight out, thus 'throwing' the first tub off the rail inches from where I lay."

At the peak of pony use in the pits, it is estimated that some 500 left the islands every year, and it is recorded that in 1878 about 200,000 ponies of many different breeds were working in the mines.

DIASPORA OF THE PONIES

Ponies for the mines were by no means the first to be exported from Shetland. The earliest recorded ponies to leave the islands were in the seventeenth century, probably bound for Holland. Nothing further was recorded until the early nineteenth century, when some went to mainland Britain, as a herd is known to have been at Ickworth in Suffolk, the seat of the Marquis of Bristol. It subsequently became the fashion for some of the great houses to have a herd of Shetlands in the park. In addition to being decorative, some were shod in leather boots designed to prevent deep hoof prints and turned loose for cutting the lawns. Before long, they were also used as children's mounts, and were put to small vehicles to be driven by ladies.

In 1885, a first batch of 75 Shetlands was exported to the United States, followed by 129 in 1887—all to Eli Elliott, the man who established the breed in America, for use as children's ponies and in harness. Since then, Shetlands have been exported in large numbers, both from Shetland and from mainland Britain, to virtually every country in the world. There are now more Shetlands in Holland, for example, than in Britain.

Those familiar with American Shetland ponies will know that they bear little resemblance to the ponies bred in the Shetland Islands, or British Shetlands worldwide. This is because the American Shetland has, over the years, been greatly changed, and now has much finer limbs, a shallower body, and a much more extravagant, hackney-like action than the British Shetland. At some stage, both Welsh and Hackney blood was introduced in some studs. The difference, from a British viewpoint, is that the British Shetland evolved with physical characteristics which enabled it to survive in its natural harsh habitat and, when domesticated, to be a working pony.

The American Shetland, on the other hand, had been developed as a show animal, with no real need for survival features. The spectacular action of the American Shetland, for instance, while brilliant in the show ring, would expend far too much energy in a winter environment where much of that precious commodity must maintain the pony's body temperature.

While it is clear that out-crossing can change a breed dramatically, it is an interesting fact that pure-bred Shetlands appear to retain virtually all their native characteristics, including their size, even after many generations of breeding away from their natural habitat. For instance, very few grow over height. This contrasts with other British native breeds, which tend to grow taller when bred outside their native environments, usually by means of a lengthening and narrowing of the long bones of the limbs. This is attributed chiefly to their being given richer feed than they received in their native surroundings. It is thought that the Shetland's small size is firmly fixed in the genes of the breed, which because of its long island isolation has probably been influenced less by outside blood down the centuries than other British pony breeds, with the probable exception of the Exmoors.

As the era of the pit ponies and the peat flittin' ponies gave way to the age of mechanization, the Shetland's role had to change yet again. More accurately, perhaps, it has had to revert to its original role as a riding pony—but with the emphasis on being the ideal riding pony for young children. So, while the majority of British-bred ponies trace back to Londonderry sires, the "Londonderry type" has been modified to some extent, making it once again rather lighter and more free-moving, to suit the many and varied demands of modern equestrian activities.

Shetlands are still bred in their native islands. However, their numbers are greatly reduced for two principal reasons. Firstly, they are obviously no longer part of the agricultural or industrial scene, and secondly, improved transport etc. has enabled studs to be set up world-wide, so that demand for island-bred ponies as stud stock has diminished. From an estimated 6,000 ponies in 1910, the figure now is believed to be nearer 600. There are said to be about 75,000 Shetlands worldwide, of which some 15,000 are in Britain.

In Shetland today, although the ponies are being bred for the "pleasure" market, the crofters' management of their stock is still largely traditional. Ponies are not stabled, and are turned out on the scattalds but, thankfully, the days of deaths from starvation are past.

Since the late 1950s, two sales of ponies have been held annually in October, one on Unst, and the other in Lerwick. The crofters bring their crop of foals, usually straight off the scattalds, often unhandled and unbroken. Buyers come from mainland Britain, and a few from overseas, mostly from Scandinavia.

Over the years, prices have fluctuated widely, mirroring changing economic conditions. In the early 1950s, trade with the United States was particularly good, and the crofters, interestingly, followed the pattern set by their ancestors and sold off much of their best stock. As before, the quality of ponies on the islands declined drastically, but this time, instead of a Lord Londonderry stepping in, a premium stallion scheme was devised by the Shetland Pony Stud-Book Society. Under this scheme, selected stallions were chosen from all over Britain, and sent to the islands to be turned out on the scattalds to run with the crofters' mares. This scheme was, overall, a great success, and is still in operation today.

One or two interesting observations have been made, underlining how quickly native instincts can be lost if not used. It was noticed that most of the stallions bred in mainland Britain, when turned out on the scattalds, had lost the instinct to search for and then round up their mares—something all Shetland-bred stallions do naturally. The mares had to be brought to them. Similarly, as they were unaccustomed to being turned out in the wide open spaces of the scattalds where they had to walk many miles a day to obtain enough of the scarce food, they lost condition during their first season very quickly.

In the early 1980s, when the bottom fell out of the pony market yet again, the crofters were once more tempted to sell their best ponies. This time, the Pony Breeders' Association of Shetland anticipated the dangers, and instituted a Filly Premium Scheme; under this arrangement, suitable three-year-old fillies were awarded premiums, provided the owners retained them for three years, thus enabling them to breed and to produce good foals. The scheme had the desired results, and very good ponies indeed are now bred in the islands.

SMALL HORSES FOR SMALL RIDERS

There is no doubt that Shetlands are ideal riding ponies for children. Their size, their willingness, their kindly temperaments, their native common sense, and the enthusiasm with which they enter into everything they are asked to do enable even tiny children to form the most wonderful partnerships with them. Until recently, however, they have tended to be regarded rather as a first pony on which the child could learn to ride, be led round a show ring, and then move on quickly to something larger. That scene has now changed dramatically, and over the last twenty years or so Shetlands in Britain have come into their own as real performance ponies, capable of competing in a wide range of equestrian events, and even beating larger ponies. This transformation began when one pony, Boffin of Transy, bred in Scotland, had a spectacular success in a Working Hunter Pony championship, over a course designed for larger animals. It included a three-foot fence with a four-foot spread, which stopped virtually all Boffin's rivals. But he cleared it with ease—a really huge obstacle for so small a pony. Shetlands had arrived as performance ponies.

Soon they were competing in show jumping, dressage, long distance, working hunter pony, cross-country, hunting and gymkhana games. A Shetland Performance Award Scheme was set up, in which ponies gained points for placing in both breed and open competitions. Ponies from the islands competed at home initially, but for a number of years, a group has come down to mainland Britain every summer, and competed throughout the country.

Most spectacular of all is the Shetland Grand National, a scaled-down version of the famous steeplechase. In this great event, young riders, colorfully dressed in racing silks and riding hats, breeches and boots, race their ponies over a circular course of small hurdles. The jockeys are all members of the Performance Award Scheme, and they have to be extremely good riders to stay on board excited Shetlands galloping absolutely flat out, and leaning over at alarming angles as they race round the bends. Falls do happen, but fortunately not very often. It is a wonderful sight, and both ponies and children love it, as do the spectators.

So popular has this event become that heats are now held at a number of shows throughout the summer, and the Grand Final takes place at the Olympia Horse Show in London at Christmas. At that, all the trappings of the "real" Grand National are included. There is a pre-race parade of runners and riders, bookmakers, a celebrity starter, a race commentator, and "owners" who are usually companies who sponsor "their" pony. Over the years, the event has raised huge sums of money for children's charities.

Eventually children outgrow the ponies for riding, but many Shetlands then change disciplines and appear in harness with adult drivers. They compete with great success in driving trials, private driving, and their real specialty, scurry driving (so

In Bressay, circa 1900, a band of ponies embarks on the first leg of the journey to England, where they will work in the coal mines of the Marquis of Londonderry.

called because the ponies "scurry"—run flat out). In the latter, where they gallop round a twisting course of cones drawing four-wheeled carriages, their agility and short-striding maneuverability enable them to beat larger ponies on many occasions.

The popularity of the breed has never been greater but, inevitably, there are some worrying features. The most serious is the extraordinary increase in the number of "miniature" Shetlands—ponies measuring 34 inches and under—being bred not just in Britain but worldwide. As long as official breed records have existed, at least since the first Stud Book published in 1891, and almost certainly before that, there have been miniature ponies in Shetland. But until the last twenty years or so, their numbers have been relatively small for the simple reason that the Shetland has always been a working pony, and a miniature pony was thus of very limited use.

There is no doubt, however, that miniatures have a role within the modern breed. They obviously give their breeders great enjoyment and satisfaction, and they can be ideal for introducing very small children to the pleasures of riding. In addition, of course, their size makes it so much easier for the youngsters to groom them, look after them, and form happy and confident relationships with them.

But it is probably true that only a minority of the miniatures now being bred are engaged in the latter activities, and the problems which have arisen due to excessive and sometimes injudicious breeding of them are real. Warnings were sounded as early as 1913, in the days when there were far fewer miniatures than there are now. Douglas wrote "It must be kept in mind that ponies of sizes less then 34 inches are of little use for practical purpose. . . . Anything that tends to make the pony merely an oddity and a toy, and to take it out of the category of useful or usable horses, is fatal to the prospect of the breed and should be resisted by breeders and judges."

Fifty years later, another great authority on the breed, Major Maurice Cox, wrote in his definitive book, *The Shetland Pony,* "It seems to me there is too much sensational enthusiasm in the breeding of these very small ponies, for the only possible use for these little ones is that of ornamental attractions. . . . Efforts to breed them smaller and smaller are disturbing. . . . What I consider to be a very great danger is the breeding of thoroughly bad little ones; it must be put on record that there are quite a number of miserable little weeds, short of bone, with poor heads, badly made, and with weak action."

Unfortunately, highly inflated prices began to be paid for

these tiny creatures—some measuring only 26 inches—by buyers from Europe and the United States. The result was that not only did some existing breeders try to cash in on this newfound wealth, but the prices attracted many who had never bred ponies before, and knew little or nothing of the very real difficulties of breeding miniatures.

Many ponies of the type described by Maurice Cox were bred and, provided they were tiny, buyers flocked to pay absurd prices for them. So concerned was the Breed Society that it introduced an inspection scheme at official sales to weed out those whose conformation was completely unacceptable. As a result, it is probably true to say that the number of really bad ponies is falling.

Nonetheless, at the beginning of the twentieth century, miniatures probably accounted for no more than about 5 percent of the breed, whereas in 2000 the percentage of mares foaling for the first time who measured under 34 inches was 60.4 percent, a figure many find alarming. After a slight decline to 57 percent in 1998 (which some hoped might be the beginning of a downward trend), this was an increase of nearly 3.5 percent over the previous year. Another fact that presents a disturbing prospect for the breed is that, in 2000, of the 47 colts out of a total of 188 which failed their inspection as potential stallions, 77.1 percent were between 30 and 34 inches. As few ponies under 38 inches are capable of being ridden in any serious way, these figures do not make good reading for those who regard all native breeds as working animals first and foremost.

The figures mean that, first, in all probability more than half of the future breed may be capable of little more than walking and trotting around a show ring in halter classes; and second, too high a number of miniature colts are unacceptable as good Shetlands. However, there is some indication that the prices for these tiny ponies are beginning to fall slightly and, if that trend continues, excessive production may also decrease. It must also be emphasized, however, that some really top-class miniatures are bred and, as always, these have a place in the breed.

Happily, the overall situation in Shetland is noticeably better. Some excellent miniatures are being bred in the islands, and they can match their larger relatives in conformation and movement. To a considerable extent this is due to the fact that virtually all the ponies in Shetland are kept under natural conditions, and thus have to walk miles each day over rough ground to find food. As a result, they tend to have much better limbs and movement than many of the more pampered southern-bred ponies.

In virtually all other aspects, Shetlands, both in the islands and in mainland Britain, are proving that they have once again adapted to the demands of the times in which they live, while retaining their essential characteristics as native ponies. Despite concerns, the overall status of the breed is very encouraging.

There is no doubt that, starting with Boffin of Transy, and continuing with the Shetland Grand National and scurry driving, together with the successes of Shetlands taking part in the Performance Scheme, the reputation of the breed as performance ponies has improved beyond recognition over recent years. There is still some way to go, but Shetlands are now in genuine contention for major awards in performance classes all over Britain. It is much rarer to hear judges making derogatory remarks such as "You can't take Shetlands seriously."

It is also vitally important for the breed as a whole that Shetlands continue to be bred in their native islands. It is an established fact that native pony breeds lose some of their hardiness and native characteristics when bred away from their natural habitat—and it is these very characteristics which have contributed so much to their popularity. Although, as mentioned earlier, Shetlands are less prone to this than virtually any other native breed, it is of supreme importance that there is a nucleus of native-bred ponies in the islands, to which breeders can and do return for infusions of true native blood, with all that implies. It is a bonus that an appreciable number of these island-bred ponies are doing so well in performance classes in mainland Britain.

Shetland ponies are unique, not just in appearance where their diminutive size, their abundance of hair in mane, tail, and forelock, and their mischievous expression make them the most easily recognized and appealing of breeds, but in their extraordinary versatility and adaptability over the centuries. There is something very special about a pony that has in turn been a mainstay of the agricultural life of an entire community, contributed hugely to the industrial success of a nation, then made the transition to being the ideal pony for the youngest rider, providing thrills and spills for older children in a variety of equestrian disciplines, and when outgrown as a riding pony, can be driven by adults at the highest level in harness events.

To quote Charles and Anne Douglas, "In the end it is idle to deny that it is not his indisputable economical validity that binds the Sheltie's lovers to him: rather it is himself—his wisdom and his courage, his companiable ways, his gay and willing service. . . . he provides . . . the dual charm of a creature at once wild and tame—wild in his strong instincts, his hardiness, and his independence—domestic in his wisdom and sweet temper, his friendly confidence in mankind, and his subtle powers of ingratiation."

66 ISLAND VIEW

Genesis

The Origins and Limits of Domestication

Modern man's great achievements do *not* include the domestication of animals. In fact, it appears that nearly all the species of animals that we recognize as domesticated were brought into the human fold by the time recorded history began. Stated another way, it was early man who tamed all the beasts that could be or would be domesticated: some birds, horses, cattle, goats, sheep, a feline that gave rise to house cats, the wolf that became hundreds of breeds of dogs.

Around the world and throughout the ages, our kind has lived near every kind of animal at one time or another. Evidence found in middens as old as Ur and in oral accounts as recent as the latest Balkan turmoil suggest that people have made nutritional use of all sorts of animals from sea cucumbers and insects (called delicacies in some parts of the world) to rodents and canids (ditto). And if we ate them, we probably tried to raise them at one time or another. Yet humans have domesticated only a few animals because the rest could not or would not be tamed, thanks to their own habits, temperaments and inherited characteristics. Those that would and could live closely with us are much in the minority.

In her authoritative textbook, *A Natural History of Domesticated Animals,* Juliet Clutton-Brock remarks that the apparent prerequisites for domestication were enumerated as long ago as 1865 by the English anthropologist and author Francis Galton. (A man of mixed credentials, he was Charles Darwin's well-traveled cousin, an admired investigator and member of the Royal Society, in short a leading scientist of his day. But he is most often remembered for inventing eugenics, a theory that has not stood time's test nor science's scrutiny, but has been used most often to justify various forms of detestable racism.) Yet Galton formulated sound ideas about the domestication of animals. Quoting these, and amplifying on them, Clutton-Brock specifies that for an animal to be domesticated, it must meet these criteria:

1. '*They should be hardy.*' The young animal has to survive removal from its own mother, probably before weaning, and adapt to a new diet, a new environment, and new conditions of temperature, humidity, infection and parasitic infestation.

2. '*They should have an inborn liking for man.*' In present-day terms this means that the behavioral structure of the species has to be allied with that of humans. It has to be a social animal whose behavioral patterns are based on a dominance hierarchy so that it will accept a human leader.

3. '*They should be comfort-loving.*' Galton meant by this that the species must not be highly adapted for

instant flight as are, for example, many members of the antelope, gazelle, and deer families. These animals will not feed or breed readily if constrained in a pen or herded too closely together.

4. '*They should be found useful to the savages*' [sic]. The primary function of captive animals to an early farming community would be as an easily-maintained source of food, an itinerant larder that can provide meat [or milk, eggs, etc.] when required.

5. '*They should breed freely.*' As Galton. . . accurately observed, this is perhaps the most necessary factor for successful domestication, as can be seen from the difficulty of maintaining breeding colonies of many species in zoos, even under the most favourable conditions that can be provided in captivity.

6. *'They should be easy to tend.'* This applies particularly to livestock animals which must be reasonably placid, versatile in their feeding habits and yet gregarious so that a herd or flock will keep together and can be easily controlled by a herdsman.

Professor Clutton-Brock concludes, "It is no wonder that so few groups of wild animals have succumbed to the process of domestication, despite more than ten thousand years of association with humans as the dominant species." Given that some animals did succumb, and lived in association with people for thousands of years, she continues, "I believe that animals bred under domestication evolve into new species, as a result of reproductive isolation from their wild progenitors combined with natural and artificial selection in association with human societies."

In other words, their association with humankind changed the animals in many ways as they became reliant or even dependent on Homo sapiens. Likewise, domesticated animals changed man's behavior. As captive flocks and herds reduced the human family's need to roam, people became more settled, cultures became more complex, civilizations more "advanced," etc.

In time our forebears must have learned—empirically at least—to breed animals differentially so as to encourage certain traits and to discourage others. Thus they raised many races of animals with disparate goals: some dogs to be aggressive guards, others to be keen-nosed trackers and still others water-loving retrievers; cattle to provide meat or milk (or both as you shall read); horses to be variously as strong as Clydesdales and fast as Thoroughbreds and sturdy as Shetlands. Thus men—even before the invention of the science of genetics or any cogent definition of "breed" or "breeding"—channeled the animals' reproductive patterns to strengthen chosen traits and suit human purposes.

Yet from the start, the process of domestication has depended not so much on man's cleverness at luring "lesser" creatures within his pale. Rather, the process of domestication has required an animal species' de facto predilection to enter our society—or at least to survive our company—and thus become a human partner. So it was in primeval Shetland, or wherever the earliest Shetlanders came from.

Several times a year, the crofters who share grazing rights on the hill or scattald hold a caa, *a round-up on foot. The people and their dogs drive the sheep to the* crö, *an enclosure built of stone, like a corral in the American West.*

Shetland Sheep

A Wealth of Wool and Words

James R. Nicolson

Shetland's recorded history starts with the Norse immigrants, who arrived in the eighth century, in the wake of the Viking explorers, and left their mark in many forms: in the place names of these islands; in our given names (and those that became surnames); in the Shetland dialect which contains many words of Norse origin; in the sheep. The Norsemen brought with them livestock, including sheep which over the centuries adapted to the harsher environment of the islands to become the famous Shetland breed that produces the world's finest wool. Theirs was a subsistence economy combining the resources of the land and sea—the basis of the crofting system that developed later.

The Shetland sheep is a small, fine-boned animal, the ewe generally without horns while the ram sports a magnificent pair of horns. Photographs taken early in the twentieth century often show a ram with two pairs of horns, a characteristic which seems to have almost died out but could appear again thanks to the infinite inventiveness and perfect memory of genes.

Like the Shetland cow and Shetland pony, the Shetland sheep is supremely at home in its native habitat. They thrive on poor vegetation which could not support a bigger breed. They are long-lived, some of exceptional stamina being known to produce lambs when fourteen or fifteen years old. They are small, hardy and extremely agile, able to outwit the shepherd and his dog in areas of rough terrain. They survive the coldest winter, protected by a thick coat of wool. They withstand periods of heavy rainfall since they have the ability to shake water off their backs—as a dog does after a shower of rain.

An unusual characteristic is the tail of the Shetland sheep, which is short, with only thirteen vertebrae, compared with twenty or more in other breeds. The tail is triangular in shape, tapering to a point, and usually covered in fine hair.

Shetland sheep come in a variety of colors—white, brown, black and shades in between. Each color has its own name in the dialect. Shades between fawn and brown are described as *moorit,* while the rare steely gray color is known as *shaela.* Many sheep are multi-colored, giving rise to a long list of adjectives to describe each combination. A *catmoget* sheep is one with a light-colored body and darker underparts; a *blaegit* sheep is white with black spots; a *bjelset* sheep has a distinctive ring around its neck, while the term *mirk-faced* applies to a white sheep with brownish spots on its face.

These natural colors became of enormous importance in the Shetland knitwear industry where multi-colored sweaters and other goods made from undyed wool became the hallmark of the fashion-conscious individual in London and Paris.

After the change from Norse rule to Scottish rule in the fifteenth century, regulations governing the keeping of sheep were laid down in the Country Acts of Shetland, legislation enacted for the government of the islands by local courts, consisting mainly of landowners. At a sitting in the Great Hall of Scalloway Castle on August 3, 1615 it was stipulated that all *dykes* or walls separating the good arable ground around human settlements from the rough hill pasture should be repaired and made stock-proof before April 15 each year, the accepted time for the start of cultivation. Sheep and other livestock were to be driven to the hills before that date.

To ensure that no livestock strayed onto the cultivated ground, travelers were required to close all *grinds* (gates) after passing through. Furthermore, it was unlawful for anyone to trespass on a neighboring *scattald* (common grazing ground) with a sheepdog unless accompanied by one or two neighbors who were regarded as "famous honest men." If this rule was broken the trespasser would be liable to a fine of six pounds in Scots money; the dog would be hanged.

The historian Alexander Fenton gives the reason for this ruling in his mammoth work *The Northern Isles: Orkney and Shetland*. He points out that the native sheep were not herded in docile flocks at that time but ran wild and had to be caught by specially trained *hadd* dogs (from the old Scots word *hald,* meaning "grasp"). These dogs ran down and held each sheep individually. For this reason it was made illegal to go through a neighboring scattald with a dog without reliable witnesses. It was also decreed that no one should keep a dog unless authorized to do so. This regulation was designed to prevent people from removing the wool from a sheep belonging to others, since ownership of the wool could not be established after it had been removed from the sheep. Chapmen (traveling merchants) who bought hides, skins or wool, had to show these to the *Bailie* of the parish, or again to two "famous and honest neighbors," who would recognize if these had been acquired illegally.

THE CROFTING SYSTEM

Sheep became a vital part of the crofting system of agriculture which prevails in Shetland (as in other parts of the Scottish Highlands and Islands) where most of the land is of poor quality and the climate harsh for much of the year. A croft is the small area of farmed land surrounding each house in a crofting community, its main crops being potatoes, oats and barley, most of which went as food for the family. Cash income was small in traditional crofting communities and most men had to find additional work which in Shetland, until quite recently, meant either fishing or seafaring, leaving much of the croft work to their wives and children.

The animals on a croft included hens, ducks, geese, a few cattle, notably a house cow, and large numbers of sheep which roamed over the wide expanse of hill and moorland which constituted the scattald. These islands had many ponies as well, but for generations they were nearly community property—like the scattald they inhabited. As Sir Walter Scott wrote in his 1821 novel, *The Pirate,* the hardy little ponies that lived on the hill were ridden or suffered to carry a burden from place to place by anyone who could catch them.

Alfred Isbister hugs a ewe, in Ham, Foula, 1938.

Although the sheep belonging to several crofters grazed together on the scattald, each family recognized their own animals since, in common with other parts of northern Europe where Norsemen left their impact, each owner had his own distinctive mark cut into one or both ears of his sheep. These earmarks were of different shapes and known as a *hole,* a *shear,* a *crook,* a *rit,* a *fedder* or a *shull.* By using both ears the possible combination of marks was increased. Each mark of ownership was recorded, as it is today, in a book kept by the clerk of the grazings committee.

An essential requirement for every scattald was the *crö,* an enclosure built of stone with a single opening, where the sheep were driven several times a year. This process was known as a *caa,* and resembled the American cowboy's roundup—on foot.

Originally there were no physical boundaries between neighboring scattalds. These were delineated on the maps of the landowners from whom the crofters rented their land, with lines drawn between prominent features, such as a large boulder left stranded at the end of the Ice Age, or a bend in a stream. In spite of the lack of a barrier between scattalds, the sheep belonging to each group of crofters generally kept to their own areas.

A change came in the mid-nineteenth century, when landowners realized that sheep of market breeds were more profitable than tenants. Large areas were "cleared" of people to make way for flocks of blackface and Cheviot sheep. The new farmers brought in experienced shepherds and sheepdogs such as the border collie to herd the sheep. The families evicted from their homes had to find vacant crofts in other parts of Shetland or join the continuing stream of emigrants to New Zealand, Canada, the United States and other countries.

It was about this time that massive stone walls marking boundaries were constructed in several parts of Shetland, dividing scattalds where for generations the sheep belonging to several crofting townships had grazed together.

The complaints of the crofters reached the ears of the government in London and led to the Napier Commission, which investigated the state of agriculture in Shetland and other parts of the Highlands and Islands. This led to the Crofters Holdings Acts of 1886, which have been described as the crofters' Magna Carta, since the legislation guaranteed security of tenure. Another improvement at this time was a compulsory regulation of the scattald to give each crofter an equal share of the livestock which the hill pasture could support.

The new rules also made it compulsory for sheep owners to "dip" their sheep annually in a chemical solution to prevent infestation by parasites. In some places this rule was adhered to with as little expense as possible, the crofters using an old boat, its seagoing life over, as a dipping tank. But before long each crö had a wooden tank properly designed and constructed for the specific purpose of dipping.

THE SEASONAL ROUTINE

The cycle started in December, when the rams, which had been kept apart from the ewes in a field or other enclosure, were taken to the scattald and set loose. The first lambs arrived in April or May and afterwards the crofters would get together on an appointed day to drive all the sheep into the *crö* so that new lambs, huddling close beside their mothers, could have a temporary mark of ownership attached—perhaps a dab of paint. At this time the best male lambs were selected for breeding, the others being castrated to be reared as *hoggs.*

Orphaned lambs were taken home to be fed from a bottle. They were known as *caddy* lambs and were generally treated as pets, much to the delight of the children in the family. (A problem was that such lambs considered themselves as part of the family and at times it was difficult to exclude them from the house.)

The arrival of the lambs coincided with the start of the growing cycle when the crofters dug or ploughed their fields to plant their crops. First of all the walls that separated the arable land from the common grazing had to be repaired and made stock-proof, to ensure that that there were no incursions of livestock to harm the growing crops.

However, some sheep, smarter than the others, always found ways to squeeze through the smallest gap in a wall or fence to get to where the grass was greener or, worse still, where the crops grew. Such sheep are called *almarks.* Their owners tried to stop their thieving habits by tying round their necks a triangular frame of wood, the theory being that this would make it impossible to get through a narrow opening. In spite of this encumbrance, such sheep still found ways of getting access to the growing crops.

The next *caa* took place in July when the sheep were penned to have their full-grown fleeces removed. At that time of year the wool of the Shetland sheep is loose and can be pulled off gently by hand without the aid of clippers, the process known as *rooing.* This is done by gently plucking bits of wool from the animal's coat until the entire fleece is removed—as it would be soon enough if the sheep were left alone to shed its old coat piecemeal in spring and early summer. (This natural shedding of the fleece bit by bit leaves the wool hanging off the sheep in clumps, which get pulled away by heather and thistles. In past days, these tufts of wool were known as *hentilagets;* many a poor widow with no sheep of her own—a sort of gleaner of wool—would collect them to knit a warm garment for herself.)

Driving the sheep towards the *crö* was a labor-intensive affair, requiring a large number of people to visit every part of the scattald. Then with much waving of arms and the swinging of jackets (taken off because of the body heat generated), the sheep would get the message and proceed in the required direction—trickles of sheep, melting into streams and eventually a river of moving backs flowing steadily towards the *crö.* Dogs were used but they were often more of a hindrance than a help because they were seldom trained for the job.

This is explained by the sheep expert Benji Hunter, an advocate of better training for both dogs and their handlers:

"In those days a *caa* was more or less organized on a communal basis. As a result, what dogs lacked in quality was more than made up for in quantity and variety, from quiet working dogs to the small types with curly tails and loud mouths. While noise did not facilitate progress, it certainly enlivened proceedings. It sometimes evoked a variety of spontaneous commands to which even the better trained dogs had not been accustomed and consequently produced little results. Not unnaturally, human as well as canine tempers became frayed. This often resulted in some short, sharp observations on the alleged errors of the over enthusiastic or the omissions of the less interested handlers, which were usually responded to with equal gusto." It is probable that these loud-mouthed, curly-tailed dogs were descendents of the old *hadd* dogs and progenitors of the modern "Shetland Sheepdog," which is classed as a working dog by the Kennel Club but is in reality an adorable pet.

In the early 1920s a group of crofters formed the Shetland Sheepdog Trials Association to improve the quality of sheepdog handling. The trials have been held annually since then, the competitions interspersed with demonstrations by handlers and their dogs from other parts of the United Kingdom.

By the middle of July the meadow grass is getting ready for cutting. Traditionally it was mown with a scythe in bunches called *swars* which were left in long rows, ready to be spread out to dry. This hay was turned constantly when the weather was sunny and then built up into small piles known as *coles* which could deflect the occasional shower. Eventually the hay was carried home and built in a large stack known as a *dess,* to be used as winter fodder for cattle. Next to ripen were the fields of oats which were cut with a scythe and bound into sheaves, placed in small *stooks* to dry and eventually built into bigger *skroos* in the yard. The last crop to mature was the potatoes, the mainstay of the family.

With the last of the crops harvested there were still large amounts of grass which had grown along the edges of the fields. This was the reward given to the sheep and young cattle which had spent the growing season outside the hill dykes. The gates were opened and sheep and cattle alike were allowed to roam freely throughout the croft, devouring the tall grass while it lasted, before returning to forage what they could find in the hills.

For the family too, autumn was a time of celebration and plenty. The men who had gone fishing or whaling returned

Might it take a village to roo a sheep? Perhaps it's just that the patient work of plucking off the wool gives the crofters of Shetland the opportunity to get together and socialize.

to the croft with their wages; there was ample food, and fresh mutton on the menu. Shetland mutton is of exceptional quality and flavor, especially in the case of sheep that have fed mainly on heather and have a high proportion of lean meat to fat. No part of the animal was wasted: The head was the basis of a nourishing broth; the entrails, washed thoroughly, were filled with oatmeal or other mixtures to make puddings and other delicacies. Autumn was also a time to prepare for winter, when older sheep removed from the flock were slaughtered, then salted in tubs of brine for later use. Many a cold-weather feast would feature *reestit* mutton, a joint that had been removed from the pickle and dried on the rafters.

In some parts of the islands these activities are still part of the ritual of autumn, in spite of the availability of canned and frozen food.

Winters are hard on the sheep as the moorland plants wither. When there is virtually nothing left on the hill to eat, the sheep depend to a large extent on the beaches, where piles of seaweed are cast ashore by storms. The sheep know when the tide is falling and head for the beaches where they forage until the tide rises again. Older Shetlanders claim that the sheep are guided by a "worm" in the *kliv* (foot) which is said to "turn" when the sea begins to ebb.

The small islands lying close to Shetland's main islands support considerable numbers of sheep during the summer when they feed on remarkably lush grass. The largest of these islands can support sheep during the winter, and here too seaweed is an important part of their diet.

Snowstorms can injure or destroy moorland sheep, which sense when a blizzard is coming and seek shelter in natural hollows and in the trenches created by peat-cutting. But such shelters give false security and it is common for sheep to become buried under several feet of snow. Their owners try to find them using poles to detect movement under the snow and bringing bundles of hay to restore vitality. When sheep are trapped under snow, it is a case of the fittest surviving as the weaker animals get trampled underfoot in the snow "cave" that develops. The fittest can survive for a surprisingly long time under snow although they are weak when the thaw comes and are often attacked by scavenging birds such as ravens and bonxies. Losses were heavy in the exceptionally hard winter of 1946–47; an estimated 30,000 sheep died in the winter of 1967–68.

In contrast to the conditions experienced in the hills, the lambs which have been selected for breeding spend their first winter in the comfort of a *lambie hoose,* a building erected for this purpose, where they are pampered, being fed regularly to ensure that they are in top condition when the time comes to join the flock.

SHETLAND'S WOOLEN INDUSTRY

Like the "dual-purpose" Shetland cattle which provide both dairy products and beef, Shetland sheep have long been a mainstay of the local economy on two counts—as the single source both of meat and of fiber. The wool from the native sheep has been the basis of a cottage industry which brought ready cash to local families. Working the wool could be done during bad weather, when outdoor work was impossible. Women and girls spent every spare moment carding the wool, spinning it into yarn and knitting plain and patterned clothes. They may not have realized that the wool of the native Shetland sheep is the finest of any produced in the British Isles; yet they certainly knew that it produced garments which could ward off the cold of a Shetland winter and which also found a ready demand in other parts of the world.

The two branches of Shetland's woolen industry—weaving and hand knitting—have their origins in the distant past, both skills having been practiced in Norse times as shown in the archaeological record of Viking settlements. During the period of Norse rule in Shetland most of the wool was woven into a coarse cloth known as *wadmel.* It was in this material that the land tax was paid to the kings of Norway. The cloth was thickened by the action of hands and feet; in many cases it was spread along the bottom of a narrow passage among the rocks on the seashore, through which the tide ebbed and flowed.

Rugs and blankets were produced locally and bedspreads, known as *taatit* rugs, were still being made at the end of the nineteenth century. The Shetland loom could produce only a narrow strip of cloth, but two lengths sewn together produced the completed ground.

While the weaving of all kinds of cloth declined, the manufacture of hand-knitted garments grew in importance. This involved not only the production of warm underclothes for the family but also garments for sale.

A great deal of work was required to prepare the raw wool for knitting. After it had been washed several times, it had to be carded, then oiled, spun and twisted into yarn. Carding—combing the wool with a pair of paddles faced with rows of fine bristle-like teeth—was a tedious process, yet when carried out by eight or ten women gathered in the *but end* of a croft house, the essential chore became a pleasant social occasion, as I relate in my book *Traditional Life in Shetland.* They sat on wooden chairs with the piles of raw wool beside them, picking it up in handfuls and spreading it out on the spikes of the lower card. They then drew the upper card over the wool two or three times before reversing the position of the cards and repeating the process. The wool was then lifted from between the spiked faces of the cards and placed between their smooth backs, where deft movements of the wrists formed it into long *rowers* or rollers ready for spinning. By the time the evening's work was finished and the menfolk arrived with their fiddles, the floor would be covered in piles of *rowers*—white, gray, moorit and black, all separated according to color. These were put away and the fiddles tuned for a dance.

In addition to the great variety of natural shades of wool, bright colors could be produced from homemade dyes. A pur-

Rooing—plucking the wool in tufts in springtime when sheep start shedding their coats—can be a family enterprise. Here John Halcrow roos an impatient animal, with Michael's help and Daisy's oversight at Cunningsburgh in 1959.

Rooing at Roermill, Northmavine: Jeemie Johnson, Mary McKenna, Maggie Blance harvest the wool from a flock of Shetland sheep, circa 1930.

ple dye produced from the lichen *Tartareus* was for long an article of commercial importance. In the same way, the lichen known as "old man's beard" produced a yellowish or reddish-brown color, while another lichen known in Shetland as *scroita* produced an orange shade.

As early as the sixteenth century there was a considerable trade between Shetland and northern Germany, as island wool was knitted into coarse stockings for export. When that market diminished there developed a considerable trade with the Dutch herring fishermen who assembled in Lerwick Harbor (then called Bressay Sound) to wait until the Feast of St. John the Baptist on June 24, the recognized date for the start of the herring fishery.

In 1837 Arthur Anderson, a Shetlander best known as the founder of the P&O shipping company, tried to popularize Shetland knitwear in London and presented some fine stockings to Queen Victoria. This was the start of a keen demand in England for fine woolen shawls, stockings, gloves and underwear. The island of Unst acquired a high reputation for its shawls of gauze-like delicacy, so light that a full-sized article nearly a yard square and weighing only two ounces could be pulled through a wedding ring. Until the 1920s the knitting of patterned garments played a very minor part in the islands' woolen industry. Then came a renewed interest in the old, traditional, multi-colored patterns which had been preserved in a few places, notably Fair Isle, which lies midway between Shetland and the Orkney Islands. A huge trade developed in pullovers, cardigans, berets and gloves, incorporating these ancient designs of Norse origin. Again the Royal Family helped to stimulate demand for this type of knitwear when in 1921 the Prince of Wales wore a Fair Isle jersey as he drove off from the first tee at St. Andrews Golf Course as captain of the Royal and Ancient Golf Club.

Knitting came to have an importance in Shetland that it had in no other part of the United Kingdom. In most parts of the country knitting was and still is a hobby, whereas in Shetland it was a vital part of the family's livelihood. Often the returns from the fishing industry were meager and the cash from the croft insignificant; but the income earned by the women from their knitwear prevented hardship. The rate of pay per hour—if anyone had taken time to make such a calculation—was very low; yet the family came to depend on this source of ready money. So the women knitted as they chatted around the fire of a winter's night; they knitted as they walked to the peat hill with a *kishie* (carrying basket) on their back for the day's supply of peats for their fire; they knitted as they walked to the local shop. A barter system developed whereby the merchant was quite happy to accept knitwear as payment for goods purchased; after all, he made a profit from both transactions.

The demand for Shetland knitwear became so great that local knitters had to hasten the process by sending their wool to mills on the Scottish mainland to be spun by machine. The practice of carding and spinning at home declined steadily in the early part of the twentieth century. But all other aspects of the art of knitting flourished, and as consumers became prepared to pay a realistic price for the garments, it became a vitally important industry—the third strand in the islands' economy along with crofting and fishing.

Many Shetlanders were forced to leave their islands in the late nineteenth century because of economic hardship. Wherever they found a new home they continued the art of knitting, often passing on to their neighbors the intricate art of Fair Isle knitting. This was the case when Jeremina Robertson emigrated to Canada in the 1880s, bringing with her a spinning wheel, knitting needles and a supply of Shetland wool. Her journey ended in British Columbia, where she met and married another Shetland emigrant, Robert Colvin. They made their home at Cowichan Bay where they cut down an area of forest to make a farm and became friendly with the local Cowichan Indians. As soon as they had cleared a few acres they bought some sheep, choosing a breed with long fine wool suitable for spinning. Jeremina knitted most of the clothes worn by her family and before long she was showing the Cowichan women how to knit thick sweaters to keep their men warm while hunting or fishing. She taught them to knit Fair Isle patterns, then as they became proficient they added their own patterns such as the "Killer Whale" and "Thunderbird." Jeremina Colvin died in 1936, relatively unknown outside that little community, yet the skills she had taught the Cowichan women became an important home industry in that part of Canada.

SAVING THE SHETLAND BREED

It is little short of miraculous that the Shetland sheep has survived as a distinct race in the islands. From a purely economic point of view, the animal's small body size is a disadvantage as it means a smaller carcass at the abattoir and less

On the hill: left to their own devices, the native sheep may thrive.

saleable meat for the butcher; crossing with larger breeds to produce a heavier animal has been going on sporadically for more than one hundred years.

Within recent years, enormous changes have taken place in the rearing of sheep. Each crofter is now entitled to apply for permission to have his share of the common grazing fenced off for his own use. In this way many thousands of acres of land which previously had grown heather and coarse grass have been ploughed, fertilized and reseeded with better strains of grass which can support larger sheep, which in turn are apt to bring higher prices in the autumn sales.

Crofting is now a highly organized activity, subject to controls laid down by both the United Kingdom's Department of Agriculture and Fisheries and more recently the European Community. It goes without saying that crofting has to pay its way and that means getting the maximum profit possible from sheep. Given the constant low demand for wool on world markets, greater attention is being paid to carcass weight. A purebred Shetland sheep, outwintered on the hills, has a live weight of 22 kg (48 pounds) on average; the offspring of a Shetland ewe and Cheviot or Suffolk *tup* (ram)—called a "first cross"—is heavier, meatier and therefore pricier. There is a keen demand from farmers on mainland Britain who purchase cross lambs at the autumn sales and fatten them on the better grassland of their part of the United Kingdom.

In spite of these changes, the purebred Shetland sheep still has an important role to play in the economy of the Shetland Islands. It is the only breed that can survive on the poorest hill land which still covers much of Shetland—and survive it does, nay it thrives in its native environment! Furthermore, only a purebred Shetland ewe can breed the valuable first cross that is so profitable; a first cross ewe produces poorer offspring. Ironically then, the rules of genetics dictate that it is in the best interests of market-conscious farmers to ensure the continued availability of little purebred Shetland ewes in order to produce the bigger, more commercially valuable, hybrid offspring.

That the Shetland breed is still in good heart is due largely to the efforts of a group of crofters led by Dr. James Bowie, who in 1927 formed the Shetland Flock Book Society. Membership is open to all crofters and farmers whose flocks conform to rigid standards regarding the physical appearance of the sheep and the quality of the wool. Shetland sheep compete in their own classes at the islands' agricultural shows and the annual lamb sales are well supported.

Likewise the Shetland knitwear industry has undergone a revolution within the last fifty years, and been partly mechanized with the introduction of knitting machines that are small enough to be installed in a knitter's home. Such a machine produces plain cardigans and jerseys which can be enhanced with the addition of a hand-knitted Fair Isle patterned yoke. The art of hand knitting is still alive and in good heart as can be seen from the great variety of Fair Isle patterned garments—gloves, scarves, hats and pullovers—knitted for one's own family and for sale in local knitwear shops. Production on a larger scale is carried on by a few local firms using machines to produce the body of a garment while the finishing touches are provided by women working at home. Recently the Shetland Knitwear Trades Association launched a campaign from its office in Lerwick which has led to increased demand for Shetland knitwear throughout the world.

Providing the raw material for this industry are the sheep, whether they be the native Shetland sheep or crossbred animals. A survey in 1994 provided an estimate of 200,000 breeding ewes in the Shetland Islands, of which 25 percent were purebred. The future of the breed would seem to be secure since these are the only sheep that can survive in the more exposed parts of the islands, and since they are uniquely required for the breeding of animals that fetch higher prices on the commercial market today. In this, a prize Shetland ewe resembles a tragic figure in the fairy tale, namely the goose that lays golden eggs. May our lamb not share that bird's famous fate, since modern man is more knowledgeable—if not wiser—than the yeoman who killed that goose and in so doing proverbially cut his own throat.

Lost Dogs, or Whither the Shetland Sheepdog?

A consensus is clear: the dog called "Shetland Sheepdog" today is not the animal that arose in Shetland, thrived there until some time in the nineteenth century when it was replaced by mainland dogs, and yet survived well into the twentieth century. While the modern Shetland sheepdog may have distinct merits, it is not the indigenous animal that inhabited the islands' crofts, richly earned its keep, and animated Shetland life for generations. More in sorrow than in anger, various people from time to time recall the erstwhile breed in hopes of finding that it survives somewhere in the world. For example, in 1999 The Shetland Times *carried the following unsigned article under the headline "Looking for a certain type of dog."*

Now living on a small farm in Wales, Anne Jackson once had a little dog. She reckons that Bijou, who was her constant companion through the late 1940s and 1950s, was one of the original breed of Shetland sheepdog.

In the 1940s, she was living in Kircudbright, and was given a puppy which had been bought in a pet shop in London as a Shetland sheepdog. She is now trying to find out whether dogs of her type still exist in Shetland.

Anne wants to make it clear that she is not looking for a pedigree Shetland sheepdog or 'Sheltie' (a mainland term) of the type developed from around the 1920s and now accepted as a breed by the Kennel Club.

In fact when she recently enquired after the original type she was told, by a leading Shetland sheepdog breeder in England, that 'we do not encourage people who want to revert to the original breed'! Which shows at least that such an older breed is widely acknowledged to have existed.

Bijou was about 14 inches high, stocky, with a very thick double coat (wooly underneath, silky on top). Her colour was a golden sable, lighter underneath but with no white, her ears and muzzle tipped black, with a wave of black hair down her back and a black ring round her tail, which was very full. Her forelegs were feathered. Her head was quite unlike that of the show breed, the forehead was broad, the eyes round, the muzzle short. She lived with no illnesses until she died at the age of 15.

In behaviour, Bijou was friendly, intelligent, interactive, always with her mistress whether she was playing, sitting still or sleeping. She would, Anne said, bring things from the farm like eggs or small chicks, holding them gently in her mouth so as not to break or injure them. She wasn't given to a great deal of barking but, when excited, she would raise her tail, though not into a tight curl. . . .

For years Anne has wondered whether her dog really was a Shetland sheepdog, since she was really not at all of the show type with its lighter body, long narrow muzzle, narrow forehead, almond eyes and different behaviour patterns.

She is now convinced that there was indeed an original breed and that Bijou was an example of it. She says she was so delighted that she can't imagine the type could totally have disappeared from Shetland, even if it may be regarded by some as just a 'coarse' small sheepdog or mongrel of no particular interest!

Whether these dogs ever worked with sheep or not is a fact at issue. There is some suggestion that they are not big enough and don't have the herding instinct. However other people point out that the original breed was stockier than the show breed and that, in any case, they used to grasp the sheep by their wool and hold them ready for rooing. . . .

Anne is. . . interested in finding a dog of the general type and character of the old breed. . . . Anyone who thinks they recognise the breed. . . or are interested, or have, or had such a dog, should drop Anne a line. The address is:

Anne Jackson, Little Wentwood Farm, Llantrisant, Usk, Gwent, Wales, NP15 1ND.

Eleven years earlier, in 1988, The Shetland Times *published letters about "early Shetland collies" or "Shetland sheepdogs," prompting a response from Dr. Stanley H.U. Bowie, a native Shetlander living in the south of England. Stanley Bowie's father was Dr. James C. Bowie, the physician who made his home at Park Hall and was highly active in the islands' farming circles. Stanley's brother Hugh Bowie was a leading crofter who kept one of the two surviving herds of Shetland cattle that enabled the breed to rebound from the specter of extinction. (A distinguished geologist, Stanley Bowie has published a number of commentaries and broadsides on Shetland animals.) Noting that his father and brother had both bred Shetland sheepdogs of "the original type" well into this century, Stanley Bowie wrote in* The Times *that the working dogs they knew on Shetland crofts*

. . . were known as *toon* dogs and were mainly used as watch dogs to keep sheep and other domesticated animals out of the enclosed arable land around a croft or farm. This they did on their own, but on occasions may have been urged by the command *sigg.*

There is no record of them having been used for 'driving, penning or catching sheep'. . . .

In 1908 the Shetland Sheepdog Club was

formed in Lerwick [doubtless Dr. Bowie's father was involved] and in the following year the breed was recognized by the Kennel Club. Breed characteristics were defined and after considerable controversy the name Shetland Sheepdog was adopted. However, in 1914 an English club was formed and rivalry broke out between breeders who wanted to maintain the original-type dog and those who wished to breed a miniature show collie. The latter variety is dominant today and it is doubtful if the original type is to be found in Shetland although examples could exist elsewhere in Britain or in America. Neither types should be confused with the so-called Shetland Collies which were once used to *caa* sheep on crofts. These were usually Border Collie crosses.

Modern Shetland Sheepdogs are very intelligent and could no doubt be trained to round up ducks, hens, geese—if that were necessary—or even sheep on a smallholding or croft, but they would be little or no use on hill land and I am certain they would make no impression on my Shetland cows.

When the editor of this book visited Agnes Leask in 1998 and 1999, she proudly demonstrated the prowess of her dog, Nan, on "the hill" where she worked a small flock of sheep. What follows is part of Mrs. Leask's exegesis, as recorded, transcribed and edited with her approval.

Hugh [Bowie] did seem to have a better grasp of what

was Shetland and what wasn't. At least with collie dogs. He condemned the Kennel Club for destroying the Shetland collie. So at least he did have a better judgment with collies.

Nan, my dog, is a border collie, but her granny, I'm pretty sure, somewhere back in her line, had Shetland collie in her. Work-wise she's taking the border collie side for working. Because her granny was a very, very keen worker. Very, very what we term "wide run"—that was when she could run away from you, she took a very wide detour to get on the far side of the sheep. Whereas, the Shetland collie, their principal instinct was more or less to drive away from you—they were more of a driving dog, as a garden dog [trained to keep browsing animals away from the gardens]. But then they were very intelligent, and anybody who had the aptitude to train them could train the Shetland collies to gather.

Basically the original Shetland sheepdog was small and mostly black and white but like everything else, you did sometimes get a different colour. And then, the lairds brought in their shepherds with the Scottish dog. I do believe that originally the Highland collie, the "Lassie" collie, was crossed with some sort of breed which is a golden breed. Now your old original Highland collie, or Scottish collie—the Lassie type of dog—was basically just the same type and shape of a dog as a border collie except it was much bigger, it was taller, heftier built. You know it had the same step in its face, a nice shape face, and the same as the Shetland collie. The old original Shetland collies had a nice step in the face.

The border collie is a working collie dog. It's smaller than the Scottish collie and black and white. Although you can get different colours. You can get white border collies, but there again that's frowned upon [by the Kennel Club]. . . .

The old, old dog in Shetland was a smaller dog still than the border collie, but not as small as the present day Shetland sheepdogs. They were mostly black. Some of them had a bit of white, some of them had black, white and a bit of tan. Black predominated. They were more of a driving dog. Their job was to stop the sheep from raiding the crops. . . .

My very first dog was partly Shetland collie. The Shetland folk called them Shetland collies but then, you see, as time evolved they got called Shetland sheepdogs. I think the Shetland people probably started calling them Shetland collies because the shepherds that were coming across from Scotland they called their dogs border collies, so the Shetland folk called their dogs Shetland collies, which was a very good way of distinguishing them.

They were all bred out. The crofters saw that the dogs the shepherds from Scotland brought in to look after the big estates were much more efficient at running out and gathering the sheep to you, rather than just being used to drive them away. And whether intentionally or accidentally, they started crossbreeding. You can imagine if somebody had one or two Shetland collie bitches and a shepherd lived nearby with a border collie dog, well what would happen when those bitches came in season. And so they started crossbreeding. And actually the crossbreeds were absolutely fantastic dogs. Most of them had the characteristics of the border collie for gathering but they had the agility and aptitude and intelligence of the Shetland collies.

The Shetland collie—the true old fashioned Shetland collie—was a very intelligent dog. It was very agile and very strong. . . . This modern Shetland collie is rather a nervy, high-strung, yappy sort of dog which has come in somewhere along the line through crossbreeding, in my opinion. . . . You start crossbreeding and you're going to end up with no end of problems [if] you just happen to get the wrong sort of genes.

A bull of noble stance and curious mien "from the Island of Shetland in the possession of John Maitland, Esq." poses for what was probably the first formal portrait of its kind. The engraving was published in 1802 by George Garrard, at his agricultural museum at Hanover Square in London.

Shetland Cattle

The Gentle Kine

Ronnie Eunson

Shetland cattle are an ancient breed which has survived the march of time and the demands of mankind, a saga of symbiosis. History reveals how man has only ever had a tenuous hold on life in Shetland, never more than a generation away from some natural vicissitude or manmade calamity, whether a hundred-year blizzard in which most of the livestock perish or the infamous Clearances of two centuries ago. Cattle have lived alongside people for as long as these islands have been inhabited, and like their keepers their continued existence has depended on a natural ability to adapt to changing circumstances.

THE ORIGINS OF SHETLAND CATTLE

The earliest settlers arrived in these islands about 5,000 years ago during the Neolithic Age from the British mainland. The origins of the first cattle are unclear and await further scientific analysis, however in light of recent archaeological investigations in Orkney and Shetland it now seems likely that they are descended from the aurochs, wild cattle which roamed the forests of Britain and Europe after the last Ice Age. Neolithic man may well have been capturing young aurochs which they tamed as best they could—no mean feat as aurochs were renowned in historic times for their ferocity and size. Archaeological remains indicate they would have stood shoulder to shoulder with the modern Charollais, originally from France but now the most numerous of the very large beef breeds on British farms.

Remains found in both Orkney and Shetland from the Neolithic period indicate a very large animal indeed. Therefore, perhaps, the first cattle of Orkney and Shetland were aurochs calves carried over the Pentland Firth from the Scottish mainland, trussed up in the bottom of a wicker-framed skin-clad boat, possibly similar to the Irish curragh in style though considerably larger.

These Neolithic cattle bones reveal interesting features of early domestication in that some clearly suffered arthritis in their joints. Were they used as draught animals? Were they kept in cramped conditions throughout the winter? Also, the teeth show many incremental light and dark ridges denoting periods of poor health: evidence, it is believed, that they lived through regular periods of great hardship—doubtless the winters. These ridges could help explain why, as time passes, the cattle bones get smaller and smaller. Quite simply, it is thought these early inhabitants could not feed their stock well enough through the winters and so two things happened: the bigger animals which needed more fodder died off during the lean times, and the people began to favour the smaller more manageable types which survived. By the Bronze Age there appear to be two types of cattle remains—larger and smaller animals. Clearly, if a theory proposed by Julie Bond of Bradford University is correct, size selection took a long time to complete.

However by the Iron Age, about 2,000 years ago, only the smaller cattle were to be found on sites in both Orkney and Shetland. At the recent archaeological dig at Upper Scalloway, the bone remains reveal an animal which would have stood just over one meter high at the shoulder. Archaeologists describe this animal as being a short-legged, stocky, short-horned beast. They feel that these animals were descended from a small gene pool, with little or no importation of fresh bloodstock, judging by a very common dental defect found among the animals, namely a missing innermost molar.

Norsemen arrived during the last part of the first millennium, many of them on the run from the King of Norway. As far as we can tell they were not the most tolerant of invaders. They introduced their own language, writing and architecture to the near-total exclusion of all else. However, throughout the Viking period, the cattle appear to remain skeletally very similar. Whether the new immigrants, being primarily a farming people, used the domesticated animals they found here or imported some of their own cattle is as yet unclear. Cattle did exist in Norway at that time which were very like those of the northern isles, and to this day the rare South and Westland breed of southwestern Norway resembles Shetland cattle closely. Without more research it is impossible to pronounce one way or the other, but it is safe to conjecture that the Norsemen would not have been able to import the large numbers of animals which they would require to establish new farms and so would have made use of the local cattle which by this time must have been very well adapted to their conditions.

Throughout the Norse rule of Shetland, cattle received little detailed description in written documents, save for their importance as a means of paying taxes to the Norse landlords, who owned much of Shetland until the sixteenth century. One common form of payment of taxes was in the "ox penny," a levy per head of cattle; another was in butter.

References in the fifteenth-century Court Books of Shetland, which survive in archives, tell of the numbers of cattle held by various landowners. Obviously some landowners held sizeable herds, large numbers of which must have been wintered outside as there appears to be no evidence of significant housing for stock. It is interesting to speculate whether these large herds of cattle were "hefted" like hill flocks of sheep, being born and bred on specific tracts of land to which they would adhere unless forced off by man or a paucity of food. Young boys or men unfit for work may well have watched the herds, ensuring that they did not break through onto the cultivated land and destroy the crops. When the landlord or one of his tenants wanted a cow suitable for milking or a beast for eating, the animal would have been selected out of the herd; in the case of an animal destined to become a house cow, she would have been broken and tamed.

It was not until 1797 that a surviving document would record detailed descriptions of the cattle themselves. The court records of Shetland contain an account of a dispute between a large landowner and one of his tenants, who at the end of the tenancy argue over the inventory of farm goods. Contained in this inventory is a full list of the cattle on the farm,

Grace Robertson milks a curious and patient Shetland cow in a field at Grannafirth, Nesting, in the early 1900s. Her milking parlor is an open field, the "hoos coo" is so docile. She wears her Sunday best and uses a shiny new pail —doubtless for the camera's benefit as photographs were still too special for every-day.

their colors, names, ages and importantly their heights. This reveals a herd remarkably consistent at least in size with earlier archaeological finds; these were still small beasts of a very similar shoulder height to those of the Iron Age.

One of the fullest descriptions of the breed is by John Shirreff in 1808. His theory regarding size was as follows "smallness is due to the scantiness of their food as neither artificial grasses, nor green crops are cultivated, nor are there any inclosures [sic] capable of protecting such crops from the multitude of sheep, cattle and horses."

At this point in history the human population was growing, thereby placing greater pressure on the already marginal farming system by forcing more people to keep more stock. Most of the records written about this time dwell on the poverty and hardship of the common people along with their dependence on a form of agriculture which was ultimately unsustainable. To explain the predicament further, Shirreff comments "little attention has been paid to improvement of any kind." This should not be seen as a denigration of the population, but a result of the social and landholding structure under which they lived.

Small farms or crofts were subdivided again and again as the numbers of potential crofters grew and this level of demand for land led to attempts to break out new crofts on the hill ground. To compound the problem, the open scattalds were unregulated, and the numbers of animals were not restricted as they are today. To eke out a living the menfolk of Shetland spent more time away from home at sea where they could earn a meager wage fishing, whaling or serving on merchant ships and leaving the womenfolk, children and elderly to work the land.

People and cattle alike had to adapt to changing circumstances; the production of milk and dairy products now became critical to the survival of the families, more so than the rearing of a calf for beef. The smaller crofts could not grow sufficient fodder to maintain small herds with bulls, therefore the priority was the milking cow. Less care was taken on the selection of a proven sire; the imperative was to get the cow back in milk to provide sustenance for the small children. John Shirreff thought these small cows were inferior to those of the western highlands of Scotland, though they produced a considerable amount of milk. The cows, presumably the milking animals, were put inside at night summer and winter.

He estimated there to be around 15,000 head of cattle in the islands at the time, a number he marveled at. He found that the natives had to graze beasts on the many *holms* or small islands around the coast in order to fatten them. Any surplus beef was salted and shipped south to Britain.

The British trend of agricultural improvement was only just appearing in Shetland by the first half of the nineteenth century as Shetland had up to then been seen as an island fishing station of little agricultural merit. Shirreff reported that "attempts made to introduce breeds of sheep from England and Scotland have been followed with the most ruinous consequences." Unfortunately these importations of stock brought a number of diseases, which once established devastated the native sheep, killing thousands.

CATTLE IN THE HISTORICAL PAST

The importance of Shetland cattle to the Shetland people cannot be overstated. Families with fathers and husbands at sea most of the time clung precariously to life with little in the way of surplus. The only consistently nutritious food young children had after they left their mothers' breast was the milk of the ubiquitous little house cow. Cattle produced the dung to give the soil fertility and oxen provided the muscle to help till the land.

A document from the estate of a Shetland laird, Gideon Gifford of Busta, illustrates that in 1771 there was a considerable trade in hiring out cows and working oxen to his tenants. At the time cows showed an average value of around £12 Scots and trained oxen as much as £27 Scots at sales at Voe and Braehoulland in Shetland. (A Scots pound was worth in the region of an English shilling.) Samuel Hibbert, writing in 1822, gives a more extensive account of Shetland cattle and their uses from evidence he amassed as he traveled from isle to isle: "These animals have long, small horns, and are of a brindled white, brown, or black colour, rarely displaying an uniform hue."

Hibbert's description of the cattle housing reveals a system which remained largely unchanged in some poorer areas until the beginning of the twentieth century. "Upon the conclusion of the ling fishery, . . . the Shetlander repairs to his scathold, and cuts down a large quantity of grass and short heath, which he spreads abroad upon the hills to dry; it is afterwards stored within the enclosure of his small farm, being piled into stacks like hay. . . . the heath is strewed along the floor of the byre, for the purpose of being well mingled with the dung that accumulates from the cows. The wet stratum is then covered over with a layer of duff mould, or dry decomposed moss, which substance in like manner, remains until it is well moistened with the dung that falls. . . . Successive strata of heather and mould, mixed with the ordure of the animal, are allowed to accumulate to a considerable height, until the pile attains such an elevation, that its removal is necessary, in order that the cattle may find sufficient head-room beneath the roof of the byre. . . . When the compost is removed, it is well blended together with a spade, and is then adapted to the land destined for cultivation." (A friend of mine tells about being sent to pick up a cow from the croft of an old lady who was no longer able to look after the animal. When he arrived the cow was looking at them through the thatched roof of the byre, having eaten through the straw thatch. Upon entering the byre, he realized the floor had risen so much with the buildup of dung and bedding that the cow was now standing up among the roof trusses.)

Recent scientific investigation into soil types down through the ages reveals that during the Neolithic and Bronze Ages not a lot of dung was applied to the cultivated land; the evidence is mainly of household waste. This does not mean that early farmers did not realize the benefits of cattle muck for the land, but that with only a small number of animals being retained close to the homesteads on a permanent basis there was not a large quantity of dung on hand to spread on the crops. During the Iron Age, agriculture became more organized, with at least some of the by-now more diminutive cows being housed. This would have necessitated the cleaning out of these houses periodically and the dung being stored in middens or dung heaps where its value would be enhanced by decomposition, thus becoming the principal source of fertilizer. Later as we enter the Norse period, seaweed too became an important element, being gathered off the beaches after winter storms. Soil samples from the crofting period show more domestic waste than dung, this being a reflection of the population increase with people living on more numerous but smaller units. These findings are not entirely as one would have expected, but they give us a picture of how climate and population pressures can affect agricultural technique.

Hibbert also explains butter making, which may seem an innocuous subject, but historically it was a contentious matter. This was due to the fact that for several centuries over half the land rents paid to the Norse and subsequent landowners were paid in butter. But the quality of this butter was notorious as being "fit for little more than for greasing cart wheels." This continued despite an Act of Parliament being drawn up to fine those who produced poor butter filled with hairs, curds or other dirt.

The traditional Shetland diet included some beef. Before salt became widely available as a preservative, the meat was air-dried in small stone buildings called *skeos* through which the wind could blow freely.

Until the early twentieth century, Shetlanders generally wore a kind of leather moccasin called a *rivlin*. This was made from the hides of cattle with the hair left on, and uncured so as to give a degree of waterproofing. Large quantities of calfskins were exported at times, indicating that numbers of calves were being slaughtered so as not to be a drain on milk supplies.

Horn was used as a material for making spoons and cups. One unusual use for entire horns was as "ludder horns," which fishermen took to sea on their boats. In fog or poor visibility, when crews were far from land, they would blow these horns to make contact with other boats. Times change, but clearly throughout history Shetlanders learned to make use of every last piece of their cattle, both living and dead.

As draught animals, oxen were the strongest available, and good ones were highly prized. They were kept for many years before eventually being slaughtered or sold.

Hibbert speaks of Shetlanders ploughing with four oxen abreast in two double yokes, which were attached to the plough by 20 feet of chain. Up till the twentieth century ploughing was done with two men, one holding the trams of the plough, and the other leading and coaxing the oxen. Usually each house kept only one ox, others being borrowed for the day's work. Apart from ploughing, oxen were used to haul carts of heavier goods like peats for fuel.

My grand uncle, who lived in Fair Isle halfway between Orkney and Shetland, often told stories about problems with oxen. Apparently an ox didn't like walking on gravel tracks because its feet became sore, being unshod. If left unattended it would gradually creep over on to the softer grassy edge of the track with the result that the cart, usually containing peats, would overturn into the ditch. Like the ponies, they did not like the heat, and so Shetlanders started their work early in the morning.

Accounts suggest that oxen and cattle in general were bigger and in better condition in Shetland than Orkney before the beginning of the nineteenth century. The Rev. Low concluded that this was due to the importation of "better" or bigger stock from across the North Sea, which more readily came to Shetland as the closest landfall and most important link for seafaring traders. However, to a visitor today it would be hard to envisage Shetland having bigger and fatter stock than Orkney's green and fertile land. The old wooden bow yokes to which the oxen were harnessed were gradually replaced during the nineteenth century by all-leather collars. These were still in use in the early twentieth century.

Since even primitive fences and dykes were expensive to erect in both time and resources, and boundaries around crofts were numerous, Shetlanders had to teach their cattle to behave on a halter and tether. They developed a kind of rope halter called a "branks," which has never really been improved

A young Shetland bull wears a well-made branks, the simple and effective native halter that stops an animal from wandering by pinching its nose when it gets to the end of its tether.

upon. It was made of two wooden cheek pieces through which homemade rope was threaded loosely. The branks was adjusted to suit whatever size of beast by varying the width between the cheek pieces. The tethering end was always left to run loose through the cheek pieces so as to control the animal. When it pulled on the tether the strain drew the hard wooden sides quickly together, nipping the animal's nose and thereby teaching it a lesson. (Sometimes unruly animals also had a half hitch turned around one ear; thankfully this practice is no longer continued.)

The tether itself had a "swill" or swivel partway along its length to hinder the rope from becoming "snooded," hopelessly twisted. At the end of the tether was a wooden or metal stake, which was driven into the ground. Tethered animals were moved quite frequently throughout the day to get to water or fresh grazing. A grand aunt of mine was said to "flit" her cows so often for their comfort that it was a wonder they had a chance to eat.

During the eighteenth and early nineteenth centuries there were a number of periods of starvation when it was documented that many cattle, sheep and ponies died. Apart from the increased pressure of population growth there was a change in the climate which led to many crop failures, and consequent shortages of winter keep.

Weakened cows led to families sometimes having to lift the cow daily to its feet in the byre in springtime. Finally, they might have had to carry her out onto the green, when the fresh growth of grass came.

Every kind of edible material was given to the cows at these times. Hibbert tells of a feed of mash made up of boiled and crushed fish bones. A kind of seaweed, called *hinniewaar,* was given sparingly either fresh or parboiled. In the byre, cows were fed little and often with small sheaves of oats followed by some root crop such as kale, small potatoes or turnips. Meadow hay was also provided, along with the chaff that remained from threshing.

In early history, while the human population was smaller, large numbers of cattle would have been outwintered partly because there were simply not enough buildings to house them, and partly because with fewer animals and a kinder climate there was more vegetation on the hills in wintertime. It is quite possible that during these earlier times the Shetland cattle were

On Fair Isle in the early 1920s, a family prepares to harrow a field for planting with at least four harrows but only one ox. Lack of draft animals did not reduce the acreage to be planted, so people themselves dragged the harrows.

of a slightly beefier type, more suited to being outwintered, since the needs of the farmers were different then. During times of dearth the cattle were sometimes bled to provide a nutritious family meal made with oatmeal and milk.

THE HOUSE COW

In more recent times, house cows were milked three times a day; at morning, at "twall" time (midday), and at night, to gain the maximum quantity of milk which could be in excess of three gallons. There appears to be no history of cheese making in Shetland within the last couple of hundred years as no recipes exist. This compares badly with Orkney where a great tradition of cheese making still goes on. An explanation of this may lie in the fact that there was just not enough surplus milk in Shetland, and so Shetlanders became more used to the byproducts of butter.

A description of butter making deserves mention:

Traditionally butter was made in a "kirn," a narrow wooden barrel, which stood about waist high, being slightly wider at the top. The milk was allowed to accumulate in the kirn for a couple of days till the wife had time to kirn using a "kirnstaff." This was a kind of plunger which was worked up and down with a twisting motion, whilst avoiding splashing the contents over the user. Red-hot "kirnin stones" were dropped into the kirn to help the butter form, or in winter the kirn was worked near the fire to the same effect. After the butter was lifted out and washed with water, it was sliced with a knife to remove any hairs. Then salt or sugar was added before it was shaped with pats.

The buttermilk which remains was called "bleddik." The curds, when strained, became "kirn milk" or "hard milk." This is rather like modern cottage cheese. A similar substance is "hung milk," although this is made from whole milk. After the curds are removed, the serum which is left was known as "bland." This was a highly prized drink, which fishermen took out to sea as a refreshment on long fishing trips. This was not the view of the Rev. Hibbert, who thought it "dangerous, causing colics, and all kinds of gripes"—despite the opinion of generations of Shetlanders. Probably the most nutritious dish was made from the first milk, colostrum, or "beest" from a newly calved cow. This was baked, becoming soft and cheesy, then sprinkled with sugar.

The increasing trend for improvement during the mid-nineteenth century and into the twentieth century saw the importation of larger mainland breeds to "improve" the native stock. The native cattle appear to have been more resilient than the sheep to new diseases from these new breeds. However, the problems for the cattle were more insidious. As demand grew in mainland Britain for beef, and bigger animals meant more income, the new "improved" breeds like the shorthorn and the Aberdeen Angus became the sires of choice. Also the credibility given to these new breeds by the government's Board of Agriculture led many people to see the old breed as just that, a relic of the past best forgotten. Farmers and crofters with drive and ambition saw that to be associated with the old breed was not to be seen to embrace the dawning of a new age of agriculture.

When the new larger sires were used on their small cows the effects were dramatic. These new cross calves grew very quickly to dwarf their mothers and, ironically, to seal their fate. Crossbreeding became the trend throughout most of Shetland by the end of the first quarter of the twentieth century.

Crofters also discovered that by retaining the first cross calves for further breeding they could produce bigger beasts. This process seemed inexorable, but to some not entirely desirable. A group of farmers and crofters, worried about the demise of their native cattle, decided in 1911 to take action. Otherwise, they realized, something precious was about to be lost.

They set up the Shetland Cattle Herd Book Society to register purebred animals and to publish a herd book annually. They also laid down a scale of points which has remained in use to this day. The preface of the first herd book ends, "These curious and handsome little creatures are so peculiarly fitted to the circumstances of their bleak and stony habitat that the utmost pains ought to be taken to preserve the breed in its purity and to improve it by judicious treatment." Some 380 cows and 39 bulls were registered in the first book by breeders from nearly all districts of Shetland.

The Shetland cow has been bred over the centuries as a "dual-purpose breed," although "multi-purpose" may be more appropriate. Shetlanders required from their cattle an ability first of all to survive the rigors of the climate and win-

Proud owners of Shetland's finest cattle prepare their animals for judging at Shetland County Agricultural Show in Lerwick around 1901.

ters of scant fodder. To counter these circumstances of winter dearth the cow relies on a natural attribute heired from her wild ancestors: when the going gets tough, they "hibernate" through the worst of the winter, an instinct probably triggered by the very short hours of daylight. They appear to live at a maintenance-only level, refusing to grow or fatten, whatever rations are offered. This instinct for survival has clearly saved the cattle, with any weaker strains dying off.

BREED PROFILE

Shetland cattle are small, with many standing around 48 inches high at the withers. They all have a very light bone structure, like Shetland sheep, with fine hair on a thin stretchy hide. Traditional standards hold that a cow's head should be small and nicely formed with a slightly dished forehead, short ears, and short fine horns.

The cow's neck is quite long and thin with the back straight to the tail root. The ribs should be well sprung with a good width between them sufficient for a couple of fingers to be inserted.

The hindquarters should be broad with a good width between the pin bones indicating good pelvic width for ease of calving.

The tail should be long with the switch virtually reaching the ground. The udder should be full, well formed, and well attached to the body with teats squarely placed and uniform.

Legs should be short with one at each corner. Hooves should be broad so that the cow does not sink when crossing soft ground. Their walking gait has the heel come down first before the toe, this being gentler on tender pasture and less liable to cut up the surface, unlike modern beef breeds which tend to stab at the ground.

Prevailing colors at the time of the first edition of the *Herd Book* in 1911 were black and dun, black and white spotted, white with blue or black colored necks, red, or red and white. Market trends dictated the prominence of black cattle and as a result today the bulk of the cattle are black or black and white with just a few reds or red and whites.

A good Shetland bull stands with most of the features of the Shetland cow, but with a strong, broad head and a powerful neck and shoulders. As the bull is half the herd, care is essential in ensuring that he is as free of faults as possible—

something the poverty-stricken crofters of the eighteenth and nineteenth centuries did not have the luxury to insist on.

THE WORK OF THE HERD BOOK SOCIETY

The subscribers of the first herd book were not only Shetlanders; a fair number came from mainland Britain, and they took effective steps to advance the popularity and stature of the Shetland breed. In 1913, these enthusiasts persuaded the Board of Agriculture to award premiums to a number of approved bulls that were retained for use in the islands. Also, the Board established a herd on its stud farm on the Scottish mainland at Inverness in recognition of the breed's valuable qualities. Soon there were classes for Shetland cattle at the prestigious Highland and Agricultural Society for Scotland Show. Apparently a few animals were even exported to Canada at this time, although there is no record of what happened to them.

By 1922, some 556 pedigreed cows were registered and bred pure, a high-water mark. But then, unfortunately, registrations declined as dissension arose among breeders over the greater value of first crosses, and the *Herd Book* ceased being published. The Society continued to meet periodically, and an attempt was made to re-establish a herd book in 1944. Though this did not happen then, the Society was able to bring pressure for a voluntary scheme of tuberculin testing, and as a consequence Shetland eventually became the first county to be declared free of bovine tuberculosis.

After the Second World War, the hand of government again caused an upsurge in crossbreeding. Authorities declined to make established incentive payments under the Hill Calf Subsidy unless the cow involved was crossed with a beef breed—and at this point in time the Shetland was not recognized as such. Somehow the breed staggered on, with the lack of suitable bulls and the shortage of semen for artificial insemination causing great difficulties for the few remaining breeders. Throughout these sorry times every kind of obstacle seemed to be placed between the remaining Shetland cattle and survival. Still, a nucleus of long-lived cows persisted on a few crofts in Shetland and at a couple of locations on the mainland of Britain, animals faithfully kept by people who believed in them.

When the Rare Breeds Survival Trust was established in 1971, Shetland cattle were placed in Category Two, only to be upgraded soon afterwards to Category One, namely "being at risk of total extinction." Five years later only 123 registrations were received by the Herd Book Society. Then in 1978, sadly, the two great stalwarts of the Society in Shetland, Mr. Hugh M. S. Bowie and Mr. T. A. U. Fraser, were forced to disperse their herds. At the eleventh hour the Shetland Islands Council stepped in with an incentive payment of £100 for every cow bred pure and £200 for every pure bull kept for service.

In 1981 the first volume of the *New Foundation Herd Book* was published which included all known and living pure bred animals. There were 82 cows registered as being born before 1981 with another 30 cows eligible for entry with only four blood lines extant. From this precarious position the Society has battled on through the persistence of its membership to maintain and promote the breed as best they could. Although still a Category One breed, i.e., critically endangered, there are now over 300 registered animals, with half of them still in Shetland.

What of the future? Has all this effort been worth it? Why go to such lengths just to save an animal that modern agriculture has largely deemed worthless? Ironically, the answers to these questions may lie in the fickle human whim of consumer fashion, which previously did so much harm in denying the breed its rightful place in the cattle world. Modern farming policy has long dictated that animal size and volume of production are two determining factors. However, there has been clearly a change in consumer desires in recent years with much greater emphasis on healthy eating, animal welfare, and environmentally friendly farming. And, as we are now learning, there is more to these little cows than meets the eye at first glance.

The Herd Book Society decided to turn to science to help explain some of the intrinsic qualities of these cattle. Using anecdotal information as a starting point, the Society commissioned research by the Scottish Agricultural College, firstly into the curious contention that cross calves off purebred cows grow at a remarkable rate, and secondly into the strange properties of the Shetland cow's milk.

An initial study proved that the calves of Shetland cows are born light but grow at an exceptional rate, showing daily live weight gains well ahead of the average commercial suckler cow. Suspecting there was something exceptional about the cow's milk, the Society once again contacted scientists at the College who agreed to analyze the Shetland cow's milk.

George Stout pauses on his way with ox and cart at Lower Stoneybrake on Fair Isle in the 1920s.

Their findings, based on a preliminary random sample taken in Shetland, showed the makeup of the milk to be somewhat different when compared with that of the standard Holstein/ Friesian cow.

It showed a much higher level of a particular fatty acid, Conjugated Linoleic Acid, which is known to play an important role in human health. CLA is credited with being an "anti-cancer agent" and also has the ability to reduce the risk of cardiovascular disease. On the basis of these findings alone, everyone who has been involved in trying to preserve Shetland cattle was vindicated; they had helped save something which may yet have a significant part to play in modern life.

Just as our cattle have had to learn to make the most of whatever they faced, so we, who farm and croft in Shetland, must identify whatever unrealized potential there exists on our islands. We hope the Shetland cow will have an important part to play in the agriculture of the new millennium. Certainly few other breeds of cattle have the credentials to equal them.

Saving the Duck

Mary Isbister and her husband raise all manner of native Shetland animals and crops at Burland Croft, their model working farm on the island of Trondra adjacent to the Mainland. Some of the plants, like the native black potato and an old barley called "bere," are nearly extinct and might have perished without the Isbisters' attention. The Shetland duck surely would have died out, a pretty iridescent black duck with a white bib. (The drake has a green sheen.) Here she writes about on their efforts.

It was in 1976 that my husband, Tommy Isbister, and I got the opportunity to take over our croft at Burland. We had always been interested in the native breeds of the Shetland Islands, but our first job was to build a new house and provide sheds, barns and fencing, etc. Then it was not long before we were looking for a Shetland cow to provide milk, butter and kirn milk (a soft cheese) for our growing family. We got quite a shock when we realized how difficult it was to find a native-bred Shetland cow.

Eventually we heard of one for sale at Walls on the west side of the island. After a rigorous interrogation by the crofter to see whether we would be suitable, we became the proud owners of a cow called Dora and her bull calf. The first years of breeding proved to be very difficult with no fertile bulls on the island; the available artificial insemination material was old and not very reliable. Eventually, after years of working along with other breeders and the Rare Breeds Survival Trust on the mainland, things began to look up.

It was apparent that not only were our native cows on the critical list, but all our indigenous poultry was almost extinct, so we began an urgent hunt to try to find native birds. The president of the Shetland Herd Book Society, Mr. Hugh Bowie, made us aware that there might still be a few purebred Shetland ducks out at the west side of the Mainland at West Burrafirth, so we wasted no time in contacting the owner. Indeed he did have ducks, although we felt their purity was in doubt because they had been running with Khaki Campbell ducks. As we felt this was as good as we were going to find, we purchased some eggs from him and bred from these, keeping the best specimens, that is the ones that appeared to be the closest to the old Shetland duck. Meanwhile, we never gave up the search for something better.

While on holiday in Foula, the island of about fifty inhabitants out on the west side of Shetland, we became aware of a single duck similar to the Shetland type. We were told that this was the last purebred duck left on the island, as the great skua had developed an appetite for fully grown domestic ducks and had taken the owner's ducks one by one. Therefore he was very keen for us to purchase this last bird before the breed died out entirely. (My husband and I both have roots that go back to Foula, and over thirty years ago I asked a very old relative how long native ducks had been on the island. She replied as far back as she could remember. This old lady was born in 1869, therefore suggesting there had been native ducks on Foula for at least 130 years, and most likely much longer than this.)

It was not long after we got the bird from Foula that we were told of another single duck, this time on the east side of Shetland, and again this was the last in the line. The owner said they had been isolated on his family croft since his grandfather's time, back in the previous century. He was also able to tell us the whereabouts of a drake he had given to a friend. His friend, realizing the importance of this matter, was more than willing to sell us this drake. So with this valuable trio, we started a breeding program.

The old Foula duck only produced six eggs in her first laying season with us, but luckily our other duck made up for it by laying about forty eggs. We stored both lots of eggs carefully until our Silky hens and bantams went broody. Then we used them to incubate these purebred Shetland eggs. The eggs were fertile and after we culled out the poorest of the drakes we were left with twenty young ducks.

At the end of this first season we became aware of a small flock of pure birds up at the very north end of mainland Shetland. This was exactly what we needed, an unrelated strain to widen the gene pool. So, immediately we exchanged a young drake with the owner. It came as a shock to learn soon after that a wild otter took all this man's remaining ducks.

We have continued to breed and distribute these ducks for sixteen years, and have done so very successfully, sending ducks as far as the south of England. There are also a few pairs that have found their way to America and Canada. We are not now too worried about a secure future for the Shetland ducks. Although the blood lines are few, there seem to be no problems with inbreeding. They have proved to be very healthy, hardy and fertile.

Although we have no scientific proof, our feelings are that these ducks are of Scandinavian origin. There was a similar type, now extinct, in the Faroe Islands; also birds of similar color but a larger size are still found in Denmark and Sweden.

Shetland hens are also on the critical list. The local research we have done points to there being two types of old hens. The first type was a bantam-sized hen, predominantly black in color, which laid a small white egg. The cockerel is colored with similarities to certain jungle fowl. We found the last remnants of these on Papa Stour, a small island just west of Mainland. The second old hen is larger in size with a tappet head—tufts of feathers on the top of the head. These hens lay a green-blue egg similar to the Chilean Araucana hen.

The older generation of folk in Shetland called these birds "galleon hens" which arouses the imagination to believe they may have arrived in Shetland off a Spanish galleon of which two from the Armada wrecked on Shetland's shores in 1588. This theory could make a link with the South American hen, which was said to have been taken from South America by the Spaniards.

Over the years of searching for our poultry, we came to fear the extinction of Shetland geese, but Mr. Bowie again came to our rescue with the knowledge that his brother and sister had taken a small flock to the Scottish mainland. We eventually traced and acquired two pair in Yorkshire and brought them back to Shetland.

Unfortunately it took years before we had much success with these birds. We think the problem arose because of inbreeding, and also because they had been reared artificially, causing them to lose their natural abilities to make a nest and sit successfully for four weeks on the eggs. Thankfully, now they have settled into their natural habitat, and the young birds are doing much better.

We have been on our croft now for nearly twenty-three years, and worked our place traditionally. But growing old native crops such as oats, bere barley, black potatoes and cabbage traditionally and organically is very labor intensive. It has not been easy but has been very rewarding. We feel privileged to have spent a lot of our lives in such a beautiful place working along with our animals and nature. Our aim has been to pass on our place in a much healthier and secure state than when we found it.

Pioneers: Far from home—an ocean away, to be precise—a pair of Shetland geese appear healthy and alert in their new domain in New York State. They may be counted among the island fowl now dispersed in several locations across North America.

Shetland Fowl

A Quest for Continuity

Nancy Kohlberg

Least known and most widely overlooked, Shetland geese and ducks embody the greatest ironies: They may descend from the oldest domesticated animals, yet they came the closest to the brink of extinction of all Shetland's surviving stocks. Still, it bears celebrating that flocks of them—like other Shetland breeds—have been established beyond the island fastness, and even in the New World.

Geese may have been among the oldest domesticates, some historians of agriculture believe. They argue that prehistoric people would have tamed first those animals that were the easiest to capture and the safest to keep or, vice versa, the easiest to keep and safest to catch. According to this theory, Stone Age people at the dawn of civilization would sensibly not have tried to tame horned cattle first, or hard-to-contain horses, or perhaps even dogs, which a consensus holds were our earliest commensals as long as 100,000 years ago. Rather, our ancestors may have first tamed birds which could be brought down with a simple weapon like a thrown stick, then rendered flightless by the pulling or breaking of a few feathers. Easily kept, such birds would not have posed much of a threat to their human keepers while they would have been a convenient source of eggs and meat as well as useful feathers, bones and oil.

In Europe the prototype of the domestic goose was the greylag, *Anser anser;* in Asia the swan goose, *Anser cygnoides.* Wherever, they have been widely welcome as farm animals since time immemorial for both the food and materials they provide and for other assets as well. For one, they are natural sentries and will raise a ruckus at the appearance of strangers or animal predators. For another, their behavior foretells the weather; in advance of severe storms Shetland geese were said to move their young into the shelter of rocks or kelp beds. The Northern Islands' geese also pleased their frugal stewards by foraging on seaweed, free fodder that is shunned by most domestic geese.

The Shetland goose, its numbers having approached the vanishing point, offers another surprise: its populations appear most numerous today not in Shetland but collectively in the gardens and farms of breeders elsewhere in Britain and North America. While there are pairs of the breed scattered around Shetland, there are now dozens of the birds in known flocks in Scotland, England, the United States and Canada. As for the comely Shetland duck, with its blue-black plumage and white bib, it could have been a boon to human health in those islands since the start of their settlement by the Stone Age wayfarers who came there in fragile boats made of animal hides.

Geese have been kept in Shetland from time out of mind though flocks of the local breed were commonly found on island crofts a century ago, now their numbers there are minuscule. Efforts to hatch and rear them in the islands in the late 1990s met with limited success at best—sometimes because of bad weather, as one spring brought late snow and hard frosts, sometimes due to apparent sterility. On one croft, a cosseted goose repeatedly failed to build a proper nest and brood her eggs—perhaps because she herself had been raised in an incubator and consequently had not experienced normal nurturing by her mother. In other words, not having been mothered naturally herself, she had not learned the skills required to be a good motherly goose. (For all our good intentions, human intervention in animal affairs is not always fruitful.)

Fortunately, earlier in the twentieth century the late Hugh Bowie exported pairs of Shetland geese to England and Scotland where the breed survived in a few scattered flocks. His brother, the geologist Stanley H.U. Bowie, FRS, wrote a breed description, which was published in *The Ark,* the journal of the Rare Breeds Survival Trust (RBST), in 1989. The salient points follow:

An unusual feature of Shetland geese is that they are sexually dichromatic, that is the genders can be told apart: the male is white (yellow as a gosling) and the female distinctly gray and white (golden yellow with a dark gray back in goslings). The goose's gray coloring is usually limited to the shoulders, secondary flight feathers, under-wing back and thigh coverts. The feathers on the back are mid-gray laced with gray-buff; the tail feathers gray with white edges and tips. Patchy gray plumage on the neck is common, with sharp delineation where the neck joins the body. Wings are well developed; when folded, the primaries cross just short of the tip of the tail. The mature goose weighs eleven to thirteen pounds, the slightly larger gander fourteen to fifteen pounds.

Latter-day breeders outside Shetland report slightly different plumage and slightly smaller birds, which proves the plain fact that any breeding stock will soon start to differentiate itself from the rest of its kind when taken from its native habitat and bred elsewhere. In a new place, the bird or animal will adapt to the new climate, different weather and unfamiliar food. It will metabolize a different array of trace elements, as these are absorbed from the soil by plants. Further, the small number of animals that colonize a new place must possess a smaller array of genes than a large population elsewhere. So, as a small breeding stock multiplies over the course of generations, the animals will become different from the original population, and more like one another since they share a smaller gene pool.

If that is true, then how could any well-meaning person act in the best interests of a breed by removing animals from their native habitat? The answer lies in the simple facts of vulnerability, or grandmother's caveat, "Never put all your eggs in one basket"—lest one accident break them all. Any locality is prone to some kind or kinds of natural calamity: fire, earthquake, flood, drought, heat, hurricane, blizzard—as is Shetland, where winter temperatures can fall to single digits Fahrenheit and triple-digit winds blow snow in barn-burying drifts. Furthermore, every organism is vulnerable to disease and groups of organisms are vulnerable to contagion. The more concentrated the flock or herd the greater the threat; the more localized an entire breed the higher the risk that any opportunistic virus or parasite could wipe it out forever. As a consequence some caring people, wise or simply sensible stewards, who have become sensitive to the value of special breeds and aware of the inevitability of risks, set out to lengthen the odds of a rare breed's survival by scattering the flock, so to speak.

What applies in terms of rarity for the largely vegetarian goose applies as well for less persnickety ducks—the handsome black Shetland duck included—which are fairly omnivorous. Thus they may have been serendipitously effective in another welcome way.

Ducks may perform a remarkable hygienic function involving the liver fluke, a parasite that causes severe diseases in mammals, such as liver rot in sheep and schistosomiasis ("river blindness") in humans along with other debilitating conditions caused by parasites. These flatworms or trematodes, which number 6,000 species worldwide, have a complex life cycle, as every biology student knows. The eggs of the adult parasite leave a host's body in its excreta. When the eggs reach a lake or marsh, they hatch into larvae, which swim about in search of a new host. Finding an appropriate snail, a larva burrows its way into the snail's liver where it changes into a sporocyst. The spore-like body in turn produces new larval forms, which leave the snail and attach themselves to objects such as blades of grass or other edible plants where they encapsulate and remain intact until swallowed by a sheep, person or other mammal, which becomes a new host. They emerge in the new digestive tract and migrate to the liver where they develop into adult flukes and cause considerable organ damage.

Ducks can break this cycle in a locality because they thrive on gastropods—snails, including those infested with liver fluke larvae. Ducks will keep a barnyard virtually free of snails and thus of liver flukes. Early Shetland folk, like Egyptians three or four millennia ago, may have originally kept waterfowl for their eggs, flesh and feathers. Yet he who kept ducks had his homestead kept relatively free of snails, and thus kept himself and his livestock free of flatworms. He and his family were healthier as a result, and likewise had healthier sheep as well. It is interesting to speculate how people we call "primitive" might have inferred a beneficial connection between keeping these ducks close and reducing the incidence of certain disabilities.

As noted in a previous section, the handsome black Shetland duck might now be extinct but for the efforts of Mary and Tommy Isbister. How encouraging it is to know that man could once domesticate an animal, and then neglect (or persecute) it to the point of extinction, and then bring it back again; all it may take is the dedication of a few members of the human race. At this writing, one may be hopeful for the continued future of both the Shetland goose and the Shetland duck. Let us be thankful for that, but not complacent; these domestic water birds already came too close to the edge.

Hoping to help ensure the survival and dispersal of these breeds, I imported a few Shetland geese and Shetland ducks to the United States in 1997 where I raised them with encouraging results and to good purpose. My husband and I had bought a farm outside of New York City a decade earlier, intending to raise and promote rare farm animals and also to do what we could to support the movement to save small family farms in America. We bought some endangered animals such as the Exmoor pony to Cabbage Hill Farm, as well as some less imperiled stocks of Highland cattle and Shetland sheep. (The latter two have recovered enough to no longer be considered "Critical" or "Rare" by the American Livestock Breeds Conservancy though they remain on the "Recovering" list.)

While on a walking tour in Shetland, we met Linda McCaig who travels there at least twice a year from her farm in the central lowlands of Scotland between Edinburgh and Glasgow where she raises Shetland cattle, Shetland sheep and a few fowl. Mrs. McCaig convinced me that the geese and ducks of Shetland were fast disappearing, and we arranged for her to ship four breeding pairs of geese and eight Shetland ducks to me. I resolved to breed these geese and ducks and to try to establish them around North America in the care of qualified people who would nurture them ably (and treasure them for their rarity as well). We had no knowledge of their genetic makeup, how closely they were related, or how old they were, and so we started a breeding program designed by D. Phillip Sponenberg and Carolyn J. Christman in *A Conservation Breeding Handbook* published by the ALBC. Bernard Gill of Canada was also very helpful and provided much useful information with regard to raising them.

The birds arrived in the United States in January 1997 and spent thirty days in quarantine. Receiving them at the farm in the beginning of February, we placed each pair in a separate run six weeks before their normal laying time in the middle of March. Only one pair laid that first year—and had only one chick, which died during a cold snap. One male died the following winter; an autopsy showed an intestinal blockage.

The following December we brought the geese into a chicken house, keeping each original pair in separate runs, since they are thought to be monogamous. We gave our widow to the most fertile pair, creating a *ménage à trois* if you will. Once the first female started sitting well, the other made a nest in another corner of the run and started laying too.

All our females laid eggs that year, 1998, about six eggs each, and most of the females were diligent sitters. Hedging our bets, we took each first clutch to incubate; when a pair laid a second clutch several weeks later we allowed them to sit and raise these young, because we prefer to raise our animals as naturally as possible. As it happened, some incubated eggs did not hatch, some brooded by their mothers did not hatch, and we had some early deaths as well. In the end, we closed the year with only six second-generation geese.

With two years experience under our belts, we improved our incubating and rearing techniques. To be specific, we kept newborn goslings in a box in the kitchen, warmed with a heat lamp; we provided chicken mash, greens and plenty of fresh water. After two to three weeks, we returned the goslings to their own parents in the chicken house and kept a heat lamp in each run in case of cold. The parents adopted them instantly and showed excellent parenting skills, being exceptionally protective. When the young were completely feathered out, and the grasses outside were greening up, we let the birds out during the day. When they became almost full grown and mature, they went down to the pond and stayed there until the following December.

In this way, I am proud to report, we successfully raised twenty goslings in 1999, and they represented as broad a genetic mix as the numbers of our original flock allowed. That year we began the process of dispersing them to interested parties in North America, sending at least two pairs each to waterfowl biologists and expert breeders in Virginia, the Pacific Northwest and Canada.

The ducks were hardly any challenge at all. We separated them into two groups, and each laid prolifically in one community nest, although the ducks never sat. However, the eggs incubated well and we had a good hatch of ducklings that survived. After a few weeks in the kitchen and six weeks in the chicken house they were ready for the duck pond where they have done well on their own. They swim all over and occasionally come up on the grass to rest and graze.

The sprightly and hardy little ducks are black with some white on the chest and a little blue feather on the wing and a tiny curl to the tail. Somewhat similar to a duck found in Sweden, they are quite capable of surviving on their own on a pond or lake and of roaming all winter if given a handful of grain each day. They are agile and alert enough to escape predators and therefore can survive without an enclosure. The females are prodigious layers, and their eggs have a high survival rate when incubated. In New York they start laying in late April and the young can be turned loose at about six weeks when they are fully feathered. They travel together in small groups and graze and rest on land occasionally. It has been a joy to watch them, and to watch visitors watch them too, because Cabbage Hill has become a model farm of sorts. (It is a site for experiments in sustainable agriculture, demonstrations in ecological management, and education projects for children and adults from around the New York metropolitan area and beyond.) So these immigrant Shetland waterfowl and their offspring have done service as ambassadors of the growing worldwide effort to preserve rare breeds.

In the past, very little has been documented about the ducks and geese of the Shetland Islands. So common as to be ubiquitous on the crofts, they were taken for granted, and consequently there are few records of their heritage and no registry of any kind. However they have existed in the islands for generations and because of their isolation some remained pure and distinct. Now, like other precious Shetland animals, a few of their growing number have left the islands and survive elsewhere, becoming changed, no doubt, but providing a kind of genetic insurance against any local calamity for the legacy of the original Shetland breed.

Island View

Lineages of Shetland

A vigorous man in the prime of life, Magnus Burgess raises Shetland cattle in Quendale toward the southern end of Mainland. He conversed with the editor a few times in the late 1990s; with his approval, the following narrative was distilled from transcriptions of those conversations.

My family has been on this croft for six generations, and I guess we go back further than that. My family was part of the Highland clearances, but we were no part of the family that went to America or wherever else. We were reseated here on this croft.

I am a Shetlander. I've been as far as Edinburgh but turned back. Three days away and I have to get back. . . .

The Shetland cow is like the Shetland ewe. She does an extremely good job, but if you crossbreed her she does an even better job. She enhances any other breed that you cross her with. So that happened of course, then the breeders got great cross-breeds that were of much more market value. It got to the point where there were very few people willing to breed and raise purebreds any more, and we were down to twenty-one Shetland cows.

Quite frankly, when you come down to twenty-one of something, it's dangerously low. That's extinct almost. So it required a miraculous effort to bring them back. It took good stockmen to remember and research what the Shetland cattle really look like, and what they were about, and why they were about, and then to get back to what they were. And that took a lot of breeding and careful handling. . . .

Shetland cattle never lost the fine bone. That was there, and their ability to produce well on very poor quality fodder. Those attributes are still with them, I am very happy to say. They're actually a dairy cow, and tremendously flexible.

If you take any species away from its homeland it changes. You take Shetland cattle to America, they'll change over a few generations. Even if you take Shetland cattle from the north of Shetland down to the southern end of the island, they'll change because they've got better grass.

Shetland cattle were never of one appearance. (She's not the prettiest cow in the world.) There were bigger ones and smaller ones depending on where they were raised and pastured. The ones that lived on the hills, they were small because that was all the hills could sustain. Just as the hill ewe is still small. . . .

The Shetland breed is probably the only breed of cattle

that has a diversity of color. (Diversity of color occurs in the Shetland sheep, the ponies, even the Shetland collies as well as the cattle.) Black is their dominant color. But in the old days, as with the Shetland sheep, a rare-colored cow was always kept just as if it was a fancy coat.

The color was largely lost many years ago because of market forces. Black-and-white Shetland calves had the buyers confused when you shipped them to Aberdeen; they thought they were small Friesians. You got £3 for that. A red-and-white one they thought was a small Ayrshire, so it was worth £2. A gray one was nothing because nobody knew what it was.

The Shetland cow evolved to be what she is during hard times—seriously hard times, when people were dying of malnutrition. If a Shetland family did not have access to a cow in the old days, some of the family would die. The old Shetland cow kept people alive.

If you go back to the days when the cows were living on rough Shetland hills, they were down to two or three gallons of milk a day. But if you had them on grass, they would reach six gallons. I would like to see the Shetland cow used in a dairy capacity because that's really where she comes from. And the Shetland cow gives quality milk.

Now the Holstein is a man-made cow, bred to exceed ten gallons, and she wears out before her time. In a Holstein herd, there'll be no cow older than five years. The Shetland cow only gets to her best at nine or ten and then she goes another ten or more. And if a Shetland cow doesn't have a calf one year, she would just carry on milking. Her ability to continue milking without having a calf surely is a thing of great importance. We have one at the moment in Shetland that I think has milked ten years without having a calf. Now you name me any other breed in the world that can do that. To sustain the old-fashioned cow, our policy is to multi-suckle—to have her foster three or four calves through the summer and have her milk all the way through to her next calving. She is bred to do that, and in turn we breed the cows that are the best milkers, so the breed continues and improves. . . .

The bloodlines go through the bull, and at one time we were down to four bulls—in the world. But we still have this diverse appearance. The fact that we have diversity of appearance means that there's a fair bit of genetic variety.

A lot of folk would like them to look more alike in appearance than I would. The cows always showed variation depending on what district of Shetland they were in. This resulted from breeders and crofters who preferred their cattle smaller or bigger or longer-legged or longer-faced. I'm sure it all came down to the crofter's preference. . . .

The whole thrust of my life has been towards production whilst maintaining, of course, the environmental assets that we have in Shetland: the diversity we've got in our breeds, in the birds and the plant life. That's what we've been about. We use very little chemical fertilizer. The total amount of chemicals used in Shetland is almost unmeasurable. If Shetland were somewhere in the middle of England, we would be called "organic."

The largest and the least horses—a Clydesdale and a Shetland—appear to scale in the authoritative book of its time, The Breeds of the Domestic Animals of the British Islands *by David Low, published in 1842. Decades before Darwin's revolutionary theory, this scientist observed "Animals become gradually adapted to the conditions in which they are placed, and many breeds have accordingly become admirably suited to the natural state of the country in which they have become acclimated." Furthermore, he wrote, "An intervention of strange blood might destroy the characters which time have imparted. . . ."*

Shetland as Laboratory

The Genetics of Island Populations

Lawrence Alderson

The diminutive Shetland pony is the smallest of all the native British ponies, and yet its neighbor on the mainland of Scotland, the Highland, is one of the largest British pony breeds, standing half as tall again. The small head of the Shetland with its concave profile, its profuse mane and variety of color, also contrasts with the heavier straight profile, normal mane and greater uniformity of the Highland. Why should two neighboring breeds be so different? They are not distantly related and they are both long-established native breeds. Could it be that the isolation of the two breeds from each other is the cause and that the evolution of the Shetland pony in its island fastness determined its distinctive characteristics?

To explore these related questions, it's appropriate first to answer a defining one: What is a breed? What are its essential elements and arbitrary parameters? The *Oxford English Dictionary* defines "Breed . . . [as] race, lineage, stock, family, strain, a line of descendants . . . distinguished by particular hereditary qualities." By pragmatic definition, animals of a breed resemble their parents and produce offspring like themselves; in a phrase, they "breed true." Further, in the lexicon of agriculture a breed is an animal race that has been modified by humankind in the process of domestication, whether that process was recent and cogently scientific or ancient and intuitive or accidental.

THE CREATION OR 'FIXING' OF BREEDS

Distinct breeds of livestock are created by the alternating effects of the mixing of different populations and the fixing of new genotypes. Military conquests and massive movements of European peoples between about 1000 B.C. and A.D. 1000 caused the spread of groups of domestic animals in Europe and laid the foundations for the broad genetic base on which breeds would be built in the following thousand years. Some breeds we recognize today appear in the earliest surviving records. The origins of White Park cattle can be traced to the Celtic fringes of Britain in the first millennium B.C.; sheep of the Soay type have been known for a similar length of time; archaeo-zoological research has identified bones in the Mendip Hills comparable to those of Exmoor ponies dating back many thousands of years; Dorking poultry were already in Britain when the Romans arrived. Most breeds, however, evolved gradually and only emerged as recognizable types in the last four centuries or so. Each local area possessed a random genetic mix (i.e., gene pool) bequeathed by the chance of history, and in many areas these became isolated populations which then became fixed into distinctive types as generation after generation reproduced their own kind without being tainted by other strains of animals from other areas.

The process of fixing a breed type is driven by selection pressures imposed by breeders (whether gentry on stud farms, modern crofters or Neolithic herdsmen), combined with adaptation to the local environment. In the case of a breed that arose "naturally"—with little or no human intervention—the adaptation factor is paramount. In any case, adaptation to the local environment is an extension of the natural process of selection of the fittest under which wild populations evolve; adaptation is a vital component of success in livestock systems. Any pressure the breeder exerts may have a contrary effect. Breeders often placed greater emphasis on obvious phenotypic characteristics, such as color or horns, than on performance. In the early twentieth century Shetland cattle were found in several colors, but by the

1960s they were black or black-and-white without exception as a result of deliberate selection. In contrast, white became the dominant color among Shetland sheep. In both cases, prevailing fashion was not dictated by performance or adaptation to the environment, but rather by the whims of the marketplace and personal preferences. This phenomenon is evident even more today in the extreme fashions of fancy fowl, the show ring for pedigreed dogs, or the obsession with withers height in cattle.

In the occurrence of some animal breeds, fashion was the dominant factor; in others, maybe sometimes because of adverse climatic conditions, native adaptation was more important. The relative influence of these two forces varies in different circumstances. Where owners of livestock are able to modify the environment, they can exert a greater artificial selection pressure. They may create an environment which enables them to breed livestock to an imaginary ideal, rather than select those animals which best suit the natural environment. In addition, they may tamper with the interaction of animals and their environment to change performance. They may mollify the effects of severe weather by providing shelter for their animals, so that hardiness becomes a less critical characteristic, or they may encourage more rapid growth by improving the quality of grazing or by providing supplementary feed, in which case thriftiness is less relevant. Under such regimes, animals grow bigger but more delicate, and transmit these qualities to succeeding generations. Animals which have lost hardiness and thriftiness become less well adapted to the natural environment of their area of origin. This scenario is seen most frequently in affluent areas. It is not typical of remote and exposed habitats—witness the relatively small size and hardiness of breeds of cattle, pony and sheep in the Shetland Islands.

Improved agricultural husbandry and animal breeding techniques in some parts of Europe in the mid-eighteenth century allowed the development of new breeds designed specifically to meet new market demands. Meanwhile the accelerating advance of technology and improved communications permitted those new breeds to rapidly influence other livestock types throughout Europe and even further afield. The genetic integrity that had accumulated gradually in the development of distinctive breeds and types was compromised in many areas, but the effect was negligible in the Shetland Islands. Whereas new breeds emerged elsewhere, old breeds persisted more strongly in their island strongholds, and this pattern was sustained even during the more frantic sequence of mixing and fixing in recent times.

A breed can be "fixed" more rapidly by the practice of mating closely related animals, i.e., inbreeding. This method, used for some time previously by breeders of racehorses, was adopted by breeders of farm livestock in the second half of the eighteenth century. One of the early exponents of the technique was Robert Bakewell, a yeoman farmer in Leicestershire in the Midlands of England. He obtained animals that suited his purpose and by mating "like to like" developed new breeds. His famous Dishley Leicester sheep were based on the gaunt native sheep of the county, interbred with others from Hereford and Lincolnshire to provide the desired missing characteristics of early maturity, blocky conformation and fine bones. Similarly, other breeders in other areas sought characteristics that they required wherever they could be found to develop different breeds; the nineteenth century in Europe, and Britain in particular, was a period when once again the genetic brew was stirred to ferment exciting new concoctions.

By the end of the nineteenth century, the urge for change and innovation in livestock breeding had subsided. The concept of genetic integrity was reasserted in the creation of stud books, herd books and flock books in which primacy of pedigree was the overriding factor. It remained so until late in the twentieth century when industrial breeding companies developed closely controlled populations of livestock, based on a narrow genetic base of one or two breeds for each species, and permitted farmers access to only the hybrid progeny of the foundation stock. As a result, the status of pure pig and poultry breeds have become seriously threatened, and other species seem destined to follow a similar pattern, so that their contribution to future biodiversity will be continuously eroded. The vulnerability of pure breeds will be exposed further by the accelerating pace of the development of biotechnology. Projects to identify genes controlling commercially desirable traits, and their precise location on specific chromosomes, are attracting a growing body of research scientists; their work will provide opportunities for genetic manipulation capable of destroying the distinctive genome and associated gene interactions of any breed. It is not likely that the dangers will be heeded in the headlong dash to insert high-performance genes into the genome of some native breeds.

The fluctuating effects of mixing and fixing are felt most keenly by breeds in areas of intensive agriculture where the demands of the marketplace are strong, as a result of either a denser human population or a richer farming environment. In Britain these conditions were found mainly in the central lowland areas. The effects were diluted with distance and lower density of population, and were least evident in the islands on the western and northern fringes of Britain.

Aboriginal breeds and types persisted most strongly in the islands and mountainous areas of the mainland. Isolation protected their integrity. The low fertility of these exposed areas demanded finely adapted animals, and the original types are small, tough and thrifty. The remnants of the old tan-faced sheep that populated much of Britain could have been seen until three or four decades ago in extinct breeds such as the Cladagh on the west coast of Eire or the Rhiw in the Llyn Peninsula of Wales, and still exist in the seaweed-eating North Ronaldsay sheep of Orkney. White Park and Highland cattle and Eriskay and Exmoor ponies are the direct and most typical descendants of the early types of cattle and pony. All are supremely adapted to their native environments. A mature

Careful rendering of a ram and ewe of "old" island breeds and a first-cross Cheviot lamb prove the appreciation of breed differences back in 1840.

North Ronaldsay ewe weighs only 15 percent as much as a Charollais ewe, but the latter would soon perish if exposed to the environment in which the former thrives. Similarly, feral herds of Exmoor ponies on the moor and White Park cattle kept outdoors throughout the year on Salisbury Plain have little in common with giant Brabancon horses or bulging Belgian Blue cattle. Charollais, Brabancons and Belgian Blues can express their rapid growth and high production in benevolent circumstances, but not in sustainable natural systems.

THE NATURE OF ISLANDS PER SE—PROS AND CONS

The coast of Britain is liberally endowed with islands on the wild western and northern boundaries, and many possess (or possessed) distinct populations of livestock. There were Cladagh sheep on the Aran Islands in Eire, Manx Loghtan sheep and Manx cats on the Isle of Man, Soay and Boreray sheep on the St. Kilda Islands, Hebridean sheep and Eriskay ponies in the Hebrides, North Ronaldsay sheep on the Orkney Islands, and a broad variety of breeds on the Shetland Islands. Indeed, lacking proof to the contrary, one could hypothesize that Shetland is the most prolific locality in the world in terms of indigenous domestic animals. Think about it: Is there another place that boasts aboriginal breeds of horse, sheep, cattle and eponymous fowl?

The Shetland Islands share some characteristics with other islands, but they also have unusual features. They sit at the outermost northern limits of the British Isles, poised uncomfortably in a harsh environment virtually as close to Norway as to the mainland of Scotland. Remoteness and a closed environment ingrain a psychology of separateness in island people, which not only resists outside influences, but also places great value on the heritage of their close-knit culture. This is expressed in the Shetland Islands not only in the variety of native Shetland breeds of livestock, but also in the associated culture which has grown up around them. Cattle enjoyed a special status in crofting communities, and close bonds developed between Shetland cows and the women who were responsible for hand-milking. The bonding was so strong in some cases that a piece of the owner's apron had to accompany a cow if she were sold; hence the term "clouty cow" (derived from "clout," an old word for a cloth or rag). The woolen industry was based on the remarkable quality of the wool of Shetland sheep. Fair Isle sweaters, named after an off-lying island, were knitted in naturally colored wool to patterns handed down through many generations, while shawls so fine that they can be drawn through a wedding ring are proud symbols of local culture in the Shetland Islands.

It is clear that there has been a strong Scandinavian influence in the Shetland Islands at least since the Viking conquest, and it has been assumed that the breeds of farm animals also owed much of their ancestry to Scandinavian livestock. This may well be true for Shetland sheep which, together with the neighboring North Ronaldsay, spring from the same root as Icelandic sheep, but it is less likely to apply to other species. Recent advances in biotechnology have permitted the origin and relationship of breeds to be defined with greater accuracy, and the results do not always confirm anecdotal evidence. Genetic distance studies for cattle based on blood types show a great distance between Scandinavian and British breeds, and Shetland cattle are embraced firmly in the latter group. Scandinavian breeds such as the Western Fjord, Telemark and Black-sided Trondheim show a closer relationship to Galloway and Belted Galloway cattle in Britain than to the Shetland. Likewise, there is no clear evidence of Scandinavian ancestry for Shetland ponies, Shetland collies or other island breeds.

Shetland, like other archipelagos, has been a relatively isolated community for much of its history, and its gene pool of livestock has demonstrated the effects of a closed population. While there are obvious great advantages in a breed's suitability to its environment, nevertheless closed populations suffer from two inherent limitations. The potential range of genetic variety is determined by the initial population. It is not possible to acquire new genetic material (that is, genes or alleles), except those that might occur by mutation, other than by introduction from other breeds by crossbreeding, i.e., introgression. The ability of the animals to thrive in their environment will be determined by the alleles which exist in the population. If some desirable alleles are missing, the population will be disadvantaged and less successful than otherwise.

To accentuate this problem, a closed population will suffer ongoing genetic erosion. Loss of genetic material occurs naturally by chance as some alleles increase their frequency at

Lambs bound for market in Aberdeen move down the quay in Quendale, Dunrossness, to a lighter that will ferry them to the freighter Norseman *lying offshore, circa 1906.*

the expense of others. The loss will be accelerated in small populations and in populations which are subjected to artificial selection based on the personal objectives of breeders. Some alleles may disappear, and this will lead to increased homozygosity or uniformity, and to higher levels of inbreeding. The population size of a breed fluctuates continually, and at extreme low levels passes through a population bottleneck, each time squeezing out genetic material that is then lost forever.

The *Shetland Cattle Herd Book,* which has had a checkered career since it was established first in 1912, has reflected several bottleneck phases. In 1922 the *Herd Book* was abandoned in acrimonious dissent and the breed returned to an unrecorded genetic wilderness; in the 1940s a second decline was precipitated when the breed did not qualify for a government assistance program, the Hill Cow Subsidy; in the 1960s the ready availability of semen from other breeds undermined pure breeding of Shetland cattle. When the *Herd Book* finally re-emerged in 1982 with the help of the Rare Breeds Survival Trust, the population had fallen to little more than one hundred animals, and the cumulative effect of recurring bottlenecks had seriously depleted the genetic base of the breed.

GENETICS AND SHETLAND CATTLE

Each characteristic and function of living creatures is controlled by genetic building blocks, known as alleles, which are positioned at one or more genes (loci) on the chromosomes. The sum of all the genes (which control the entire organism) is known as the genome. Different alleles at each locus exert a different effect, and this provides the variation that is necessary for selection and adaptation. At the main locus which controls color in Shetland cattle, one allele produces black and another produces red. During the twentieth century the "red" allele reached a very low frequency and for many years no red calves were born. Red could easily have disappeared, especially during one of the bottleneck phases, reducing the color variation in the breed and leaving only the "black" allele at the color locus. However, because red is recessive to black, it was carried by chance through generations of breeding hidden in the genome of black cattle, and finally red calves reappeared in the final quarter of the century.

The same principle applies to other characteristics, and genetic erosion will reduce variability for performance or functional efficiency as it does for color. The desire of breeders to select for uniformity within a breed, often realized through the imposition of strict breed standards (as after the establishment of a herd book society), encourages homozygosity and results in the loss of genetic variability. Intense selection may create a degree of homozygosity that becomes detrimental, and is likely to be accentuated in small populations.

The combination of these factors may determine the viability of a population. A small population closed to outside influence, especially when subjected to the combined effects of selection for uniformity and the chance loss of genetic material, will experience significant inbreeding. If the genome contains "defective" alleles, and they become concentrated, extreme inbreeding could lead to extinction. However, if the genome is free from deleterious alleles, or if they exist at a very low frequency, then a population may be able to tolerate higher levels of inbreeding. It is likely that many small populations have become extinct, but those that have thrived may owe their survival, in part at least, to their tolerance of inbreeding and to a fortuitous array of genetic traits that serve individual animals well in the given environment. Thus these individuals prosper and produce young most like themselves, with those selfsame suitable traits which in turn enable them to survive and reproduce. The breed of Shetland cattle probably falls into this category. Until the last quarter of the twentieth century it was confined largely to its native islands. There may have been some introgression, but not enough to deserve serious reference, and of little consequence in comparison with the small genetic base created by the dominance of two bloodlines, the Glebe and Heather, from the 1960s onwards. In such a population inbreeding is inevitable, and in individual animals it may be extreme, yet the incidence of defects is low and there is little evidence of depression of performance or viability as a result of inbreeding.

Shetland cattle in general have remained true to type because they have evolved from within, rather than being changed by external genetic influence or inept human interference, say, in breeding decisions based on fashion. Although crofting no longer determines the system of management for most Shetland cattle, they still are kept in non-intensive systems which demand the same qualities of native adaptability. Even when a major herd, the Knocknagael herd, was established on the mainland of Scotland, it occupied exposed mountain pasture above Inverness. Friesian bulls were imported to the Islands in the 1920s, but they could not tolerate the local conditions; genetic distance studies indicate that their long-term effect has been negligible. No doubt Shetland cattle have lost some genetic variability because they have been maintained as a small, closed population, with the chance that important alleles may have disappeared; yet it is remarkably fortunate that a special and distinctive allele has remained, one that apparently gives the milk of the breed anticarcinogenic properties. Fortune has smiled on Shetland cattle in this instance.

A DIASPORA OF SHEEP

The evolution of Shetland sheep stands in marked contrast to that of Shetland cattle. Although there is similarity insofar as both breeds were contained mainly within the Shetland Islands until the mid-twentieth century, thereafter their paths diverged. Many Shetland sheep were exported from the Islands. New flocks were established, not only throughout the United Kingdom but also in other parts of the world. Genetic variability was maintained by the rapid increase in the numbers of the breed, and was aided by the relatively large ratio of rams to ewes. The creation of new flocks, often in the hands of hobby farmers with a few acres, and the relative ease of keeping a ram, enabled breeders to avoid significant levels of inbreeding. The distance between flocks resulting from geographic spread also prevented the dominance of any one line or flock, and the contrasting objectives of the Island breeders and those in the remainder of the United Kingdom reinforced this advantage. While the flocks in the Islands were influenced by the demands of the British Wool Marketing Board for the production of white wool, flocks elsewhere were founded with the specific objective of conserving the variety of natural colors found in Shetland wool.

The benefit of greater genetic variability conferred by these developments must be weighed against the potential disadvantages. As a breed, Shetland sheep are now found in widely contrasting locations. Some may resemble the native Islands in some respects, but others provide a very different environment. The flocks in each location will evolve to suit both the conditions and the whims or breeding objectives of the flock owners in that area. Therefore the type of the sheep may increasingly depart from the original type. Even worse, breeders in some areas may not have the same commitment to the native Islands culture which embraced the true breed standard, and may succumb to the temptation to indulge in creative breed development through introgression.

Even the Shetland Islands themselves have not been immune to introgression. There have been significant introductions of Cheviot sheep which have mingled with some Shetland sheep flocks. After the two British Friesian bulls were imported in the 1920s, some unkind commentators described Shetland cattle as miniature Friesians. Thus the remoteness of the Shetland Islands has not proved an absolute barrier to the invasion of foreign breeds, but the genetic integrity of both Shetland cattle and Shetland sheep has been protected by two other lines of defense. Imported breeds were not adapted to the local environment; the climate, grazing, mineral balances, and management all favored native breeds. Local fervor also resisted the conflicting temptation of speculation, and it was these forces that ensured the ongoing purity of the native breeds.

SUSTAINING RARE BREEDS—A BLUEPRINT

A feature of the livestock industry in the developed world in the second half of the twentieth century was the emergence of understanding the importance of genetic conservation, and of the need to link each conservation project to a proper habitat or environment. There were early isolated symptoms of this awakening in legislation in Eire to protect Kerry cattle, and in the Texas Longhorn projects in North America, but in general the movement started slowly, first in Europe—particularly in Britain and Hungary—before gathering pace as it spread globally. At the end of the century there were more than twenty-five national organizations under the coordinating umbrella of Rare Breeds International. The philosophy that they preach and support has been of immense benefit to island populations. Islands are recognized as special locations that often possess distinctive types of livestock. Thus the attention of the Rare Breeds Survival Trust in Britain was turned at an early stage to the conservation of seaweed-eating North Ronaldsay sheep from the Orkney Islands. Similarly, in America and New Zealand, Ossabaw pigs and Arapawa sheep respectively were the subject of emergency rescue projects. These three breeds or types serve to distinguish the need to evaluate each island population on its own merits. While North Ronaldsay sheep are an ancient native breed universally accepted as a vital genetic resource, both Ossabaw pigs and Arapawa sheep are relatively recent feral populations which possess distinctive characteristics but whose relevance to their habitat excited much debate and dispute.

Shetland Island breeds fall into the category of long-established native breeds. They have special characteristics and are adapted to their area of origin. Shetland ponies were too numerous to fall within the remit of the Rare Breeds Survival Trust, and neither dogs nor poultry attracted the Trust's interest, but Shetland sheep and especially Shetland cattle have been prominent players on the rare breeds stage, and the latter provide an excellent example of the application of genetic theory and breeding programs to effect the conservation of typical island breeds.

In the late 1970s there were only 111 Shetland cows in less than 30 herds, and 25 bulls in sire lines tracing back to four foundation bulls, Glebe Wallace, Heather Marshall, Knocknagael Tommy and Knocknagael Donald. Genetic variability was maintained by limiting inbreeding through the maintenance of each sire line, and semen was collected from bulls representing all four lines. However, it was realized at an early stage that herds established in richer pastures in Scotland and England, or even the desire to follow current fashion as typified by the importation of large beef animals from the mainland of Europe, might result in selection for greater size. The appearance of one or two untypically large bulls, such as Araclett Heracles, Tanyard Taurus and Westerhouse Dandy, confirmed this suspicion. Thus bulls selected as candidate donors for the Semen Bank were measured to ensure that they remained true to the Shetland type and size. This policy, although somewhat crude and unsophisticated in its early form, observed the two fundamental requirements of a conservation program, namely the maintenance of genetic variability and the selection of animals true to the original type of the breed.

Subsequently, more effective techniques were developed. Analyses of complete pedigrees replaced reliance on sire line charts; calculations of coefficients of inbreeding were computerized; linear assessment procedures gave a more precise picture of the style and conformation of breeds and individual animals; finally, analyses of the contribution made by founder animals to the genotype of the breed and individuals provided a conservation tool for animal genetic resources that proved invaluable in designing breeding policy and mating programs. How did this apply to Shetland cattle? A prominent sire, Stanemore Odin, provides a good example.

Under the old sire line system, Odin appears to be a dominant influence. Out of fourteen bulls available for artificial insemination in 1998 through the RBST Semen Bank, five direct male descendants of Odin joined him on the stud: Ash Dougall (linebred grandson), Waterloo Charlie (grandson), St. Trinians Mansie (grandson), Garths Gunnar (great-grandson), St. Trinians Lawrie (great-grandson). Two others, Galfrid Ashley (great-grandson) and Darose Dewi (great-grandson), were descendants through daughters and granddaughters of Odin. Further influential descendants, such as Troswick Beach (son), were used through natural service. But use of full pedigree analysis reveals that the influence of Stanemore Odin, while significant, in reality is much more in balance, and the foundation bull of his sire line (Knocknagael Donald) has not made the largest ancestral contribution to the breed. In general it seems that the breed has maintained a wide genetic base, but a note of caution must be sounded. The assumption is based on available evidence, and account must be taken that pedigree information has been available only since the herd book was established in 1919, that some information is missing, and that some information may be incorrect.

It is assumed that foundation animals in a herd book are unrelated, but in practice they may be closely related and it is probable that this was so for Shetland cattle. The truth is hidden in the unrecorded pedigrees of pre-herd book animals. In some breeds, records of registered animals indicate a remarkably narrow genetic base. In the first half of the twentieth century a White Park bull, Faygate Brace, contributed about 40 percent of the ancestry of his breed, while a descendent, Whipsnade 201, similarly contributed more than 30 percent to the ancestry of the breed in the second half of the century. This is an extreme example which by good fortune has had excellent results; it might so easily have been different if other bulls had been dominant. Despite the success of the White Park, prudence dictates that genetic conservation policies should be designed to prevent other rare breeds falling into a similar pattern.

Two pairs of horns was a common trait among Shetland rams until the mid-20th century. Now they appear much less frequently in Shetland flocks.

The dangers of inbreeding notwithstanding, it is a practice that should not be rejected out of hand. It is not in dispute that the concentration of a defect or undesirable characteristic through inbreeding can be detrimental to a breed. A Shetland bull, Wykham Gentian, was heavily inbred and it is possible that the undesirable temperament of this bull and some of his descendants was exposed through inbreeding. On the other hand, inbreeding can be used to maintain distinct families within a breed, and this can be part of a constructive and helpful breeding policy. Where inbreeding is used in a planned program to concentrate the qualities of a specific renowned ancestor, it is known as linebreeding. The example of the mating of half-sib progeny of Knocknagael Roy (Jarl and Morag II) to produce Stanemore Odin, who in turn was the double grandfather of Ash Dougall, demonstrates the successful use of this technique. Despite its clear advantages, linebreeding of such severity will cause some genetic erosion. It might have been preferable if it had been less intense, but if several families within a breed are maintained in a similar way, each linebred to selected animals within its own founder stock, the basis has been created for an effective program of genetic conservation.

Various models of rotational mating and cyclic crossing programs have been devised for the maintenance of genetic variability in a population, but all rely on the basic principles of the system which I developed in 1974 for use with endangered breeds. Breeding animals in a breed are divided into groups on the basis of close relatedness, similarity of type, or both. The system will work with as few as three groups, but a minimum of six groups is desirable. Some females within each group are mated to males of the same group to maintain the particular characteristics of that group. The remainder of the females in the group are mated to males from the previous group. All females always remain in the group of their birth. This system combines linebreeding and cyclic crossing in a manner which offers the best opportunity to maximize genetic variability and minimize inbreeding in the population as a whole.

The speed of genetic loss is greater in small populations, and the potential value of rotational mating systems is enhanced in such circumstances. Island breeds in their native habitat are necessarily small populations, and are ideal candidates for the application of rotational mating systems and other carefully monitored controls. Herein lies a potentially obstructive conflict. The fervent independence of island peoples is not readily compatible with the imposition of programs which impinge on their freedom of action and decision-making. The genetic erosion that may take place before the gravity of a situation overcomes the reluctance to surrender a degree of individual liberty may in some cases pass the point of no return for the conservation of a genuine native island breed. The native breeds of the Shetland Islands have survived many and varied threats in the past. They have been sustained by their isolation and by the local fervor of the islanders. Their ability to maintain their genetic integrity in the future will be sorely tested as human and animal populations become more mobile, and as modern biotechnology reinforces the desire of those owners who wish to experiment and manipulate the genotypes of their animals.

Despite the gradual awakening to the realities of genetic erosion in the closing years of the second millennium, a sustained assault from several quarters continues to deplete the genetic heritage embodied in our native breeds of livestock. The global genetic reservoir is leaking profusely. Its genes are siphoned off by the depredation of big business, either seeking speculative short-term gain or locking up the values of special local adaptation by patent. Its genetic purity is polluted by the profit ethic that strives to maximize output per animal by the application of extreme management and the most advanced reproductive technologies. What a recipe for disaster—satisfying immediate demands of a few at the expense of long-term needs of all. In a world of fast foods and destruction of finite resources, the heritage of genetic biodiversity looks increasingly fragile as species, breeds and varieties follow an inexorable path to extinction.

The vulnerability of valuable genetic resources in current circumstances prioritizes their need for a refuge, and remote areas or islands often offer a haven where local breeds enjoy an enhanced chance to survive. Their survival may be a potent factor for our future well-being. When high-octane intensive agricultural systems run out of gas, we may need to rely on the example of native adapted breeds, in tune with the environment, to bring the livestock industry back to its senses and start to redevelop sustainable systems of production. The thrifty qualities of Shetland livestock, ignored by mainstream agriculture for so long, may well command increasing attention and respect.

Island View

An Older Breed of Sheep in Shetland?

Agnes Leask, a life long crofter, has served as president of the Shetland delegation to the Scottish Crofters Union. In interviews edited for publication here, she prides herself on raising a kind of sheep which she believes to be older than the official Shetland breed, the animal described in the flock book of 1927.

I was six years old when I had my first sheep of my own. It was one of a batch my mother had got at a local sheep sale and because it was different from the rest—black with a star on its body—I asked if I could have it. And that is really what some of my stock progressively originated from. The first one, she produced a lamb when she was a year old, and that was kept. The first grandchild was really pretty. It was black with a white head, black circles around its eyes, quick breast, quick forelegs, white socks around the back legs and a white tail. And of course it was called "Beauty." And all down the years, even after I left and went away to work, my mother, every year when an old one was being culled, she'd mark a young one of that same line for me.

When I got married and we got this croft where we now live, my mother gave back to me four sheep for the in-bye. . . . And that year there came a terrible blizzard and it wiped out all the stock except three. They were Shetland sheep. Real old Shetland sheep. Otherwise they wouldn't have got the colour and the mixture of colours.

The old originals have evolved over the centuries to thrive and live in the conditions here. I don't agree with the flock book of 1927. It's an engineered breed. You won't get modern flock book sheep surviving in the hills without a bit of pamper. That's speaking from experience. We did try a few flock book ones some years ago, because, let's face it, the old breed Shetland sheep are very poor commercial value, ...

The older type sheep is smaller. Because of a smaller body size you get a little bit less fleece off it. Because of the smaller body size and the type of sheep they are, they're slower maturing. They're leaner. They don't put on the same depth of meat that the flock book ones do. If you put them on very good grazing, oh yes they put on condition, but then you've got a greater layer of fat to meat. Now market demand is for lean meat. And if you're going to use the older ones you have to leave them a year longer to mature to get the depth of meat on them.

Way out on the hill I keep the Shetland. Pure Shetland. In the old Shetland you get all mixtures of colours. And this was one thing that I disagreed with the flock book. The only colour that the flock book would originally recognize was *moorit* which is the browny colour, and white. They would not recognise any other colours. Their recognising other colours and accepting them into the flock has only come in the last five years or so. With the result, you see, that they were totally breeding out the colour. They were breeding out the originality of the breed.

As to the breeding practices when I was a child, a crofter would have a white ram for two years, then a moorit ram, then a black ram or a grey ram. He would vary it. You'd only ever have a ram for two years among the sheep, because if you kept him for any longer he was going back to his own daughters and granddaughters. So it was a two-year cycle. And then you got the different colour. You changed colours. And this is what gave you the different colours of the sheep, and that gave you the multicolours as well.

Our hill sheep only come in for three weeks in December to the run. That's all that they're ever on the croft. It doesn't matter how severe the winter weather is, they're on the hill. Plus the lambs from that season. The lambs are born in May and the female lambs, we keep them as replacements. They go to the ram in December and the lambs are born in May. They're born on the hill, under natural conditions. That sheep has no feed. Whereas we see with neighbours who have flock book animals, come November they take their female lambs off the hill and they are fed on the in-bye with concentrates all winter. . . .

It was 1927 when the Shetland was established as a breed. Obviously the breeders had been crossbreeding for a year or two with Merino to establish the breed. Now the idea behind it was to give a slightly bigger animal and, as you know, Merino wool is beautiful wool. They realised that if they crossed that with the Shetland they'd hopefully get a uniform fleece but a good quality fleece, which they did. I'm not condemning them for what they did. In fact they did an excellent job. But then it's an animal that is not really suited to the very poor quality hills. You can have them on the hills yes, but at vital times in their life they've got to have a little mollycoddling. They've got to come in and get concentrates before they lamb, to give the lamb that bit more strength and to give them that bit more milk. . . .

I keep the old Shetland sheep for the simple reason that if that dies out it can never more be revived. And I do feel that the older type animals have very important places to play in the future. I've always had this—I don't know what you'd call it—obsession to preserve the best from the past, and of course Shetland sheep was one of my priorities.

Older animals are more disease resistant, they can live without pampering, they don't have to have antibiotics to keep them alive. They don't need high protein feeds to keep them functioning. In this day and age, in particular, I do think it would be a healthier better world if farmers and smallholders and crofters went back to the native animals of that area they live in. They wouldn't have the cost of input into them, they would be producing healthier food for the housewife. I feel very wary of buying lamb from the butcher's shop, if we didn't have our own lamb. Because you honestly don't know what it has been fed on and what has been injected into it in its short life. If you're going to rear an animal on antibiotics, high protein feed, what residues of that is in the meat? What effect will that have on the human in years to come?

The only known image of a putative "Shetland Pig," found in a Victorian travelogue, shows a feisty critter tethered to a table leg in a croft house, where he seems quite at home.

The Missing Pig

Beyond the Brink

Richard H. L. Lutwyche

We can be certain of one thing: there is no Shetland breed of pig today. The question remains, was there ever one? Certainly there is proof of pigs in the Shetland Islands, both in terms of fossil finds and folklore. There is also ample evidence that pigs and pig meat were never very popular in Scotland and the further north one went, the less so. So let us investigate the breed that was or was not—in any case the Shetland denizen that now is no more.

First, consider some reasons that militate against the existence of a Shetland breed of pig, reasons involving both sound practicality and native prejudice.

Life on the Shetlands was always tough, especially in winter when just maintaining cattle was difficult; years ago many cows would have died before the first days of spring arrived to give new life to nature's greenery. Likewise, sheep and ponies often perished. Yet pigs are not grazing animals; they have a digestive system very much like man's. They do not readily convert the cellulose of grass and therefore, in many ways, they compete with man for the food they need. They will eat cereals and vegetables, and if driven by hunger can break down all but the strongest barriers with their muscular neck and shoulders to find nutrition. Pigs roaming these islands would have been a menace to anyone trying to grow crops of corn or potatoes or storing them in winter.

Pigs can and will eke out a living foraging on roots and insects, acorns and beech mast, hedgerow fruits and berries and even carrion, but it was not by accident that in times past the traditional month for slaughtering your pigs was November, because by then the "free feed" or "pannage" ran out and any flesh your pig had put on was soon taken off again in helping the animal simply survive. On islands with few trees, let alone forests, slaughtering time may well have been even earlier.

On islands, pigs have been known to forage on the foreshore for small crabs and shellfish as well as for carrion washed up on the beach. Pigs are intelligent and in times of hardship can find meager pickings where others fear to tread. Before sanitation, pigs were the main waste disposal units in most towns and would have been reduced to such searching on the Shetlands.

So keeping pigs on these islands would have been a struggle and one which most people would not have bothered with had not the flesh been so important to aiding the survival of the family during the privations of a harsh winter. Moreover, the Celts of Scotland were never very keen on pigs and largely shunned both their husbandry and the delights of their flesh throughout much of history. There has always been a certain uneasiness between the Scotsman and what he called "the grumphy" which made the swine a rarity there anyway.

This taboo was exacerbated even further in the fishing ports around the Scottish coast and throughout the far-flung islands (although, as others have mentioned, many Shetlanders do not count themselves as true Scots). Nonetheless, the superstitions were, and indeed are, as widely held there as anywhere. Superstitions built up over generations made the pig one of the most feared and reviled creatures. Fisher folk in northern Britain will not mention the word "pig" whilst at sea as they believe it will bring disaster upon their heads and their boats. Instead, if reference to it is somehow essential, they will refer to "the thing," "old grumphy," "the Grecian," "curlie tail," "guffey," etc. Much of this has to do with the old belief that "pigs see the wind," and therefore have a major influence on the weather, which could mean life or death for the boatmen.

This habit arose from ancient Nordic mythology where giant boars were deemed to be the powers behind storms. It was not helped by Biblical tales such as that of the Gadarene Swine where the devil was seen to enter pigs through the hole in the animal's foot. Should the dreaded word slip out, then the fishermen immediately touched a ring bolt—"cold iron"—as an antidote. This magic was so strong that when the same folk were in church and the story of the Gadarene Swine was retold, there would be a murmur around the kirk of "cauld airn" as the men touched the steel on the soles or heels of their boots.

If a pig crossed their path on the way to the harbor, fishermen would turn round and cancel the day's fishing as some disaster would be bound to occur if they proceeded, as modern scholars like Christine Chandler and Peter Haining have written. Other words were also taboo, for different reasons: salmon, cat, fox, knife, salt, among them. But "pig" evoked the strongest of superstitions. In a community like the Shetlanders', where fishing was an essential activity, it seems unlikely that pigs would have been at all popular.

Yet despite all the evidence against pigs thriving in such circumstances, the hardy and resourceful Shetlanders would have appreciated many of the pig's virtues, such as its ability to convert waste materials to valuable protein which, when cured with salt and smoke, could be kept over winter. The pig's prolificacy and fast growth rates made it a much better bet as a meat producer than sheep or cattle; this very fact is supposed to have stopped the ancient Chinese from converting to Islam.

We know there were pigs in Shetland. Not only do living people tell of them, but a teacher, Jill Blackadder, recently unearthed some bones of island pigs at Breckon near Cullivow. These remains reinforce the stories handed down that the pigs were small, about 90 or so pounds in weight, about the size of a Labrador dog. They were hairy, with a distinct mane, and had small curved tusks about the size of a man's little finger. The colors, as far as one can tell from local lore, were various ranging from bluish gray, spotted, to brown and white. The closest to them in appearance may be the feral pigs of Ossabaw Island, near Savannah, Georgia, in America, which were supposed to have been descended from Spanish swine that survived shipwrecks, supposedly as pigs survived the Spanish Armada galleons that were wrecked in the Shetlands. But to me it seems more likely that what we know of these Shetland swine and the Ossabaw's appearance shows how primitive western pigs turn out when left to their own devices and become feral. Consider the 1870s writings of H. D. Richardson:

> In the Orkney Islands, the Hebrides, and the Shetland or Zedland Islands, there exists a small and very peculiar breed of swine. In size this hog is remarkably diminutive, scarcely equaling a good-sized terrier dog in stature. [The single extant nineteenth-century picture of a pig in Shetland shows such an animal.] Its color is grey, its coat coarse and bristly. Dr Hibbert [an authority of a half century earlier] calls it 'a little, ugly, brindled monster, an epitome of the wild boar, yet scarcely larger than an English terrier,' and thus draws a graphic sketch of this strange little swine's character and habits:
>
> 'This lordling of the Shetland scatholds and arable lands ranges undisturbed over his free desmenses; and, in quest of the roots of plants, or of earthworms, hollows out deep furrows and trenches in the best pastures; destroys, in his progress, all the nests which he can find, of plovers, curlews, or chalders; bivouacs in some potato field, which he rarely quits until he has excavated a ditch large enough to bury within it a dozen fellow commoners of his own size and weight. Nor is the reign of this petty tyrant altogether bloodless: when a young lamb is just dropped, it is then that he foams, and, as Blackmore has pompously sung, 'it flourishes his ivory war,' never quitting his ground till the grass is stained with the red slaughter of his victim.
>
> 'These little swine are uncared for by their proprietors, and left to shift wholly for themselves. They know no shelter, save such as they are fortunate enough to find beneath a whin [sic] bush, or under the shelter of some friendly rock or bank. They know no feeding, save such as their own ingenuity enables them to procure; yet notwithstanding all this apparent privation (for the localities they inhabit are, as the reader must be aware, none of the most fertile or abundant in such food as would

> accidently [sic] fall in a pig's way), they are by no means deficient in flesh, especially in autumn, when they are said to be in the best and highest condition. If driven home, and put up to feed, they fatten on inexpensive food, with considerable rapidity, and increase also in actual size, so as to astonish a person previously unacquainted with them.'

Of Richardson's contemporaries, William Youatt is remarkably similar in his descriptions and David Low reinforces the picture painted although in briefer terms. Robert Henderson admits to not really knowing of the type of pigs in the northern Scottish island groups. W. C. L. Martin quotes from Sir Walter Scott's *Journal of a Voyage to the Shetland and Orkney Isles*: These hogs "are an excellent breed—queer-looking wild creatures with heads like wild boars, but making capital bacon." He goes on to describe how the young piglets were a frequent quarry of the eagles on the islands.

Alexander Fenton, the modern authority, tells us that there were large numbers in the Shetlands in the early nineteenth century but that by 1936 there were just 94 and by 1969, so few that the British Department of Agriculture had ceased recording them. Further, he reports that the old native type had died out by the 1920s and that pigs since then have been brought in from Aberdeen stock markets making them similar to the breeds and types found throughout Britain.

The colors of the original island pig, he tells us, were black or dark red, brown, dirty white and tawny and that they had "short arched backs and flat bodies, covered with long stiff bristles over a fleece of coarse wool." He also reminds us that the Shetlanders, like the Orcadians, would make strong ropes from this fiber which were used for descending the cliffs to obtain eggs from the nesting sea birds. Such ropes were said to be stronger than those made of hemp. He completes his description by saying that they had "strong, erect ears and strong noses." He also records that during the middle of the nineteenth century, there was some "improvement" brought about by means of swine imported by some of the Greenland ships.

Now consider reports from people on the islands today. It seems that pigs kept on the islands were at least partially feral. Like other livestock in these islands—only more so—the pigs one hears about locally were left to their own devices and lived "on the hill." Letting them run wild would have partly countered the arguments above about competing for human food; feral pigs were left to sink or swim. As the cows of Shetland got very little feed and care from crofters, and as the ponies and sheep got less if any, what pigs there were got none at all, or what might be called negative care. Thus their scavenging skills were put to good use.

A lady who has lived on in the Shetlands all her life recalls being told as a child before World War II about such pigs by an elderly aunt. In her aunt's childhood, one of the tasks given to croft children was to keep an eye on the local "wild" pig during springtime to discover where she made her nest for farrowing. (All swine, given the opportunity, will build a nest of straw or bracken or some such before giving birth.) Once the piglets were a few weeks old and able to be sustained away from the sow, the children watched the sow until she left the litter to go foraging, at which time they would snatch a piglet. It was then taken home and raised for bacon. Crofters could only take piglets from swine which lived on their part of the common ground on the hill, and could only take one suckling so that future breeding was not inhibited.

This last point seems to relate to ancient laws on the Shetlands according to Dr. Hibbert's 1822 *Account of the Shetland Isles*, "that none have more swine than effeiring [proportional] to their land labouring," and "that none have [i.e., may allow] swine pasturing on their neighbours' lands, under the penalty of ten pounds besides damages." According to William V.S. Youatt in 1847, these laws were largely disregarded, and as a result, the pigs were allowed to wander where they would, doing damage as they went.

In many instances, the snatched pig would be reared for bacon, living in the house with the family, tethered by one leg, very much in the way they were in America's Appalachia and Ireland, where it was said they were often better cared for than the children. Pig meat cured and/or smoked would have been one of the mainstays of protein for the family throughout the winter. Pork takes salt much more readily than other meat and the mutton and seabirds would not have lasted so well (nor, perhaps, have tasted so good) as home-cured hams and bacon.

An islander, who has conserved old Shetland breeds on his croft and researched the Shetland pig as much as anyone, believes that these pigs were still around at the beginning of the Second World War. His grandfather died in 1946 and used to refer to them as if they had been quite commonplace up until the recent past. However, by then they were not popular since they damaged the wildlife by raiding and destroying the nests of ground-nesting birds. Their rooting also did damage to crops and gardens.

So, clearly, a distinct type of pig once dwelled on the Shetland Islands. Was it a breed? I think not; at least not in the terms we might define one today. It is easy to borrow the definition of "a breed" from one of the authoritative dictionaries and say that it is a population of animals who bear offspring that resemble themselves—that "breed true to type." But this does not really tell us how breeds of farm animals came about. Until breeders sent out to improve livestock—men like Bakewell and Coke in England in the eighteenth and nineteenth centuries—started plying their skills, there was no real perception of trying to breed true to type. Thus local types remained in place largely through inertia. Rudimentary improvement might begin if someone did introduce new genes by spotting that a relative or acquaintance had stock of a different and better type, acquiring an example and crossing with his stock. In most instances, when "breeds" started to evolve through more considered experimentation and selection, it generally happened in cattle, sheep and horses. These were the creatures that fine gentlemen and wealthy landowners associated with. The pig was more a utility creature, very useful for waste disposal, prolific and fast growing, but associated more with the peasant. This was, after all, the creature that the working man kept to survive on. No nobleman in those times would really associate himself with the lowly swine.

Thus pig improvement did not really begin until the early nineteenth century and the evolution of breeds did not start until the middle of that century and then it was still largely undertaken by the common man. The dictionary will tell us that a breed is defined when one reaches the stage of the offspring breeding true to both their parents—that the visual recognition is fixed in each generation. But to my mind, it is more than simply that. A breed is something that has been selected rather than accidentally evolved. All the old breeds of Britain, which used to be known as the stockyard of the world, came about this way. The Hereford and Aberdeen Angus cattle; the Suffolk and Leicester sheep; Shire and Clydesdale horses. All were developed for a purpose, to meet the needs of the men who developed them until they became quite distinctive.

From the evidence I have seen, there is little or nothing to support this happening in the Shetland pig. Our contemporary observers indicate that it was very similar to the pigs found in the other island groups off Scotland. The account from earlier in this century of young pigs being snatched to raise for bacon refers to "wild pigs." In other words, there had been no development of the Shetland pig. It was the result of environment on its ancestor, the wild boar, and I wager that if examples of those pigs had been removed from that environment and taken to breed in mainland Britain—where they would forage in areas of plenty—they would have developed after several generations quite differently and the dictionary definition would no longer apply.

Some readers will doubtless complain that my definition would deprive many island breeds of recognition, but I disagree. The sheep breeds from the Scottish islands, whilst all having some similar characteristics, are all sufficiently different to be readily recognizable. Thus the Hebridean is different from the Soay, which is different from the Boreray (Hebrides), and all are different from the North Ronaldsay of the Orkneys, and the Shetland again is quite different. Shetland cattle differ from those found elsewhere, but there is little from contemporary records to suggest that the small Shetland pig was in any way particularly different from other feral swine in those northern British parts. This is not to denigrate it. From what we have learned it was perfectly adapted to survive the harsh environment and a more highly bred animal would have found that much harder to do.

Also, as I said at the beginning, they exist no more. And that, perhaps, is the greatest shame. We can now only have the fading memories and brief accounts in books mainly over a century old, as no specifically agricultural work published since mentions them to the best of my knowledge. This means that the people of the Shetlands have lost part of their past and the rest of us are the poorer for it.

APPENDIX

Editor's Afterword and Acknowledgments

The reader will have inferred that this compendium is meant to be an informing introduction to a remote place and its animals whose special value derives in part from their rarity. Sad to say, that community appears all the more precious today after the outbreak of foot-and-mouth disease in Europe in 2001, a contagion that blessedly did not reach Shetland. Might these distinct Shetland breeds have been saved by the very remoteness that allowed them to become distinct? Might some parochial trait or gene make them less susceptible to that plague? In either case, might remote Shetland be a biological bank that provides living capital for the decimated farms of Britain—after this emergency or perhaps one that is yet to come? Whatever the answers to those questions, this little volume is intended to introduce far-flung readers to the particular wonders of the Northern Islands as well as to some unique complexities of this remarkable community.

All that said, I thank the people of Shetland for enabling this book, particularly those whom I was fortunate to meet on three visits to their extraordinary islands. The native contributors of chapters and sidebars are of course James Nicolson, Ronnie Eunson, Mary Isbister, Agnes Leask and Magnus Burgess. Yet others gave of their time, knowledge and experience as well: crofters, curators, publicans and hostlers, a bureaucrat, a scholar, a journalist, mariners and musicians, et al. Those authors aside, I appreciate the contributions and consultations of their many neighbors, including: the venerable crofter Tommy Fraser, one of those who actually saved the cattle breed; John J. Graham, author, educator, anthologist and lexicographer; Andrew Harmsworth, then the Shetland Council's agriculture expert; Jim Johnson, chairman of the Shetland Herd Book Society; Mrs. J. Evelyn Leask, secretary of the Society; Jim Smith and his sister Eva, celebrated pony breeders of Berry Farm; the owners and staff of *The Shetland Times.*

For yeoman service at the Shetland Museum in Lerwick, I am especially grateful to Ian Tait, for making the museum's visual resources available across an ocean. Dealing with the unforgiving logistics of archival recovery and transatlantic transmission, he was both tireless and unfailingly good-humored—a personification of Shetland vitality.

Going farther afield, I thank the other British writers: Valerie Russell of Caithness; Richard Lutwyche and Lawrence Alderson, both then of the Rare Breeds Survival Trust in Warwickshire; Andro Linklater, the Orcadian who returned to the more remote archipelago and wrote its history. In Edinburgh, I benefited from the guidance of the librarians of the National Museum of Scotland and the Royal (Dick) Veterinary College at the University of Edinburgh. In Somerset I visited Dr. Stanley H. U. Bowie, FRS and his hospitable wife.

In Scotland's Central Lowlands, I met Linda McCaig, a sheep and cattle breeder and regular visitor to the Islands, who first lofted the notion of a personal book about Shetland's native breeds and one person's experiences with them. After that project came to naught, Posterity Press developed this more ambitious plan: to find qualified experts who would write with objectivity and authority about each of Shetland's storied animals, then to combine their essays with others on related subjects into this omnibus, this anthology of Shetland and her denizens, human and animal.

My colleague Nancy Kohlberg is the sole American author of a chapter here. Her expert viewpoint derives not from being a native champion of a Shetland breed but rather a foreigner, a participant in the dispersal of an imperiled animal and a recipient. She was also a prime catalyst of the book, as it was she who first fielded the idea of a volume about Shetland's animals, introduced Mrs. McCaig to me, and then later as my co-editor partnered in the evolution and the development of this complex book from its conception to final edit.

Elsewhere in America, the volume benefited variously from people who performed certain tasks and sundry deeds or who provided inspiring insights and/or sound advice: In

Blacksburg, Virginia, Professor D. Phillip Sponenberg of the Virginia-Maryland Regional College of Veterinary Medicine; in Corvallis, Oregon, David Holderread of the Holderread Waterfowl Preservation Center; in New Haven, Connecticut, the staff of the Yale Center for British Art; in Greenwich, Connecticut, the venerable polymath A. C. Viebranz; in Williamsburg, Virginia, Richard Nichols, director of Coach and Livestock at Colonial Williamsburg.

Two organizations provided important assistance in both particular advice and general encouragement, Britain's aforementioned Rare Breeds Survival Trust, and the American Livestock Breeds Conservancy in Pittsboro, North Carolina. Champions of rare breeds, these organizations serve them directly, as through RBST's program of marketing the meat of rare livestock breeds through retail outlets, thus providing a practical economic incentive for farmers to continue raising these animals. (Mind you, if there were no commercial market for such animals, few farmers could afford to continue to breed and raise them. The future of the vast majority of rare or "minor" livestock breeds depends on there being viable commercial demand for them.) They also serve the cause indirectly, as through the ALBC's informing publications and periodicals, the best forum and sounding board of the American rare breeds community, indeed its nexus.

For Posterity Press, I thank them all.

Philip Kopper
PUBLISHER AND EDITOR

Authors

LAWRENCE ALDERSON

Recognized on both sides of the Atlantic as an authority on livestock genetics, Lawrence Alderson may be best known for his decade-long service as executive director of Britain's Rare Breeds Survival Trust in the 1990s. A member of the working group that established the RBST, he has a master's degree from Cambridge (as well as trophies for boxing and debating). Born to a farming family that raised pedigreed livestock, Alderson has written scores of scientific papers and a dozen books including *The Chance to Survive,* which is now regarded as the standard text on rare breeds conservation, and *A Breed of Distinction,* a monograph on White Park cattle.

Founder of Rare Breeds International and a trustee of several associations of breeders and conservationists, he serves as a director of both the Traditional Breeds Meat Marketing Company in Britain and the Kelmscott Rare Breeds Foundation in the United States. A master of both the art of breeding and the science of genetics, he is the creator of a new high performance ovine breed, British Milksheep, which have been exported to several countries.

RONNIE EUNSON

Born in Lerwick, the son of a local businessman and a librarian, Ronnie Eunson attended the Anderson High School there, then entered Edinburgh University where he earned a bachelor of arts degree in English language and linguistics, whence his knowledge of Celtic languages and Old Norse. During nine years in the booming construction trades (thanks to the prosperity brought by the Sullom Voe oil depot), he and his wife, Anne, gathered a nest egg and bought their first croft, "seeing crofting as a much more satisfying lifestyle. Both our families have a tradition in crofting stretching back as far as our families' histories are known."

While raising sheep has been his main task, he long ago bought a Shetland cow and now has a herd of twenty-three breeding cows and bulls of two distinct bloodlines. "We assess all progeny, retaining those we like until they are mature when we decide whether to breed from them. Stock not retained are grown on and fatted for sale through specialist butchers." Mr. Eunson has served as president of the Shetland Cattle Herd Book Society, vice president of the local branch of the National Farmers Union of Scotland, and chairman of the

Shetland Crofting, Farming and Wildlife Advisory Group, the county's principal conservation organization.

ANDRO LINKLATER

The noted travel and history writer Andro Linklater belongs to a distinguished literary and political clan of Orcadian origins that includes his late father, the author Eric Linklater, his mother, a local political leader, and his brother, a leading Edinburgh journalist. Andro Linklater was born in Edinburgh and educated at Oxford.

His latest book is a World War II history and romantic thriller, *The Code of Love: The True Story of Two Lovers Torn Apart by the War That Brought Them Together.* His oeuvre includes the spectacular adventure *Wild People, Travels with Borneo's Headhunters;* the biographies *An Unhusbanded Life: Charlotte Despard, Suffragette, Socialist, and Sinn Feiner,* and *Compton Mackenzie: a Life.* He completed a history of the Black Watch regiment which his father had left unfinished and he has written a book for children.

NANCY KOHLBERG

An active supporter of the rare breeds movement in the United States, Nancy Kohlberg is the proprietor of Cabbage Hill Farm in Mt. Kisco, New York, an exurb of New York City. The farm, her home, is a sanctuary for rare breeds of domestic animals and fowl and a living laboratory for experiments in sustainable agriculture. For example, its demonstration projects have included raising tilapia, a food fish, recycling the tank water through hydroponic beds of herbs and vegetables, then selling both the vegetables and fish to local specialty markets. Cabbage Hill Farm has also supported the conversion of a disused railroad station into a farm market and organic cafe.

PHILIP KOPPER

Once a journalist, then an eclectic author, Philip Kopper became the publisher of Posterity Press in 1995. Educated at Yale, he was a reporter for *The Baltimore Sun,* a performing arts critic at *The Washington Post,* a magazine editor, then the director of publications at the National Endowment for the Arts. His own books include a natural science perambulation, *The Wild Edge, Life and Lore of the Great Atlantic Beaches;* a survey of pre-Columbian North America, *The Smithsonian Book of North American Indians;* and histories of three museums, *The National Museum of Natural History; The National Gallery of Art, Gift to the Nation,* and *Colonial Williamsburg.*

RICHARD H. L. LUTWYCHE

Richard Lutwyche has always lived on a farm, and raising livestock—pigs in particular—has always been one of his vocations. Pioneering the economic revival of small farmers has been another, as in 1991 he established a mail order business dealing solely in the meat of rare breeds. This initiative led the Rare Breeds Survival Trust to create its larger program three years later, the burgeoning Traditional Breeds Meat Marketing Scheme, which now involves many parties throughout Britain. Thus breeders, stockmen, abattoirs, butchers, specialty shops and consumer groups are involved in advancing the popularity of meats derived from rare breeds.

In 1996 Lutwyche joined the staff of the RBST as publicity officer and editor of the Trust's journal, *The Ark.* (In addition, he is now development manager of the meat marketing program.) Extramurally, he has written for magazines including *Country Living, Country Life, Pig Farming,* and *Countryside,* and he recently completed a comprehensive book on the pig, its history, biology, utility, etc. An authority on British breeds of pigs, and a leader in the movement to conserve rare livestock breeds, he is founder and secretary of the Gloucestershire Old Spots Pig Breeders' Club and a council member of the British Lop Pig Society.

JAMES R. NICOLSON

The son of a crofter-seaman, James R. Nicolson was born on a croft on the Mainland of Shetland near Aith. After primary school in Scalloway, where his family moved when he was a child, he attended secondary school in Lerwick, then left the island for Aberdeen University where he studied geology, earning a combined master's and bachelor of science degree. He worked for five years surveying mineral deposits in Sierra Leone, West Africa, returned to Shetland "to satisfy an earlier ambition to become a commercial fisherman," then turned to writing.

For the past twenty years Nicolson has edited the monthly magazine *Shetland Life.* In addition, he has written ten books, including *Traditional Life in Shetland, Shetland Folklore,* and *Shetland Fishermen*, his latest. In 1990 he revised Eric Linklater's classic *Orkney and Shetland, An Historical, Geographical, Social and Scenic Survey* for publication in its fifth edition.

VALERIE RUSSELL

Valerie Russell, the journalist and author, is an authority on Britain's breeds of native ponies. Her books include *British Native Ponies* and *Shetland Ponies;* her popular columns and features appear in such periodicals as the weekly *Horse and Hound,* and the *Scottish Equestrian.* She also writes for *Scottish Farmer* and *Farmers' Guardian.* Born and schooled in Australia, she came to England and worked in London theater, then pursued interests in biology, took a bachelor's degree in botany and zoology at the University of London, and turned to writing.

Becoming a freelance specializing in equestrian matters, she moved to the New Forest where she found herself "surrounded by New Forest ponies." She found them and their ecology so interesting that she embarked on her first book, *New Forest Ponies* published by David & Charles, which appeared in 1976. A long-time resident of Caithness in the north of Scotland, she is now finishing a book on Highland ponies for Whittet Books.

Bibliography

Alderson, Lawrence, *The Chance to Survive,* Pilkington Press Ltd.and Rare Breeds Survival Trust, Northampton (1994, third edition).

Alderson, Lawrence and Bodo, Imre (editors), *Genetic Conservation of Domestic Livestock,* CAAB International, Wallingford (1992).

Bowie, S.H.U., *Shetland Sheep* (pamphlet), Lerwick (1994).

Bowie, S.H.U., *Shetland Wool* (pamphlet), Crewkerne (1994).

Chandler, Christine, *Every Man's Book of Superstitions,* AR Mowbray & Co. Ltd, London (1970).

Clutton-Brock, Juliet, *A Natural History of Domesticated Animals,* Cambridge University Press, Cambridge (1999, second edition).

Edmonston, A., *View of the Ancient and Present State of the Zetland Islands,* Edinburgh (1908).

Fenton, Alexander, *The Northern Isles—Orkney and Shetland,* John Donald Publishers Ltd., Edinburgh (1978).

Graham, John J. and Laurence I. (editors) *A Shetland Anthology, Poetry from Earliest Times to the Present Day*, Shetland Publishing Company, Lerwick (1998).

Haining, Peter, *Superstitions,* Sidgwick & Jackson, London (1979).

Henderson, Robert, *A Treatise on the Breeding of Swine,* Archibald Allardice, Leith, Edinburgh (1814).

Hibbert, Samuel, *Account of the Shetland Islands,* Hurst Robinson & Co., London (1822).

Hill, Douglas, *Magic and Superstition,* The Hamlyn Publishing Group Ltd., London (1968).

Howarth, David, *The Shetland Bus,* Shetland Times Ltd., Lerwick (1998).

Linklater, Eric, *Orkney and Shetland, An Historical, Geographical, Social and Scenic Survey,* Robert Hale, London (1990, fifth edition revised by James R. Nicolson).

Martin W.C.L., *The Pig,* revised and edited by Samuel Sidney, George Routledge & Sons, London (1857).

Nicolson, James R., *Shetland,* Newton Abbot (1972).

Nicolson, James R., *Traditional Life in Shetland,* London (1978).

Radford, E. and M.A., *Encyclopaedia of Superstitions,* Hutchinson & Co. (Publishers) Ltd., London (1969).

Richardson, H.D., *The Pig—Its Origin and Varieties,* Frederick Warne & Co., London (ca. 1872).

Scott, Sir Walter, *The Pirate,* foreword by Andrew Wawn, Shetland Times Ltd., Lerwick, 1996.

Shirreff, John, *Agriculture in Orkney and Shetland,* 1808.

Sinclair, Sir John, Bart., *A Statistical Account of Shetland, 1791 to 1799 (Drawn up from the Communications of the Ministries of the Different Parishes)*, Shetland News, Lerwick, 1925.

Youatt, William, V.S., *The Pig,* Cradock & Co., London (1847).

Picture Credits

All illustrations—except those specified below—were provided by the Shetland Museum in Lerwick, and may be found in its picture archives. The exceptions include:

Front cover: Shetland Ram, painting by Robin Hill

Back cover: Scalloway Castle 150 years ago, watercolor by John Christian Schetky (1778–1874) from his travelogue, *Sketches and Notes of a Cruise in Scotch Waters on Board His Grace the Duke of Rutland's Yacht Resolution* (London 1849). Yale Center for British Art, Paul Mellon collection.

Title page: photo by Timothy Kopper.

Page 96: Shetland Bull, a handcolored engraving by George Garrand, 1802. Special Collections, Edinburgh University Library.

Page 134: Shetland Pony and Clydesdale, from *The Breeds of the Domestic Animals of the British Islands: The Horse and the Ox,* by David Low, hand-colored engraving by Fairland (after drawings by M. W. Nicholson and paintings by Mr Shiels), London 1842. Yale Center for British Art, Paul Mellon Collection.

Page 156: Engraving of a croft interior, with tethered pig, *Picturesque Scenes of the Shetland Islands,* London, 1890.

Shetland Breeds, 'Little Animals. . . Very Full of Spirit'—
Ancient, Endangered & Adaptable
was designed by Kathleen Sims
at Sims Design Co., LLC in Washington, D.C.

The type throughout is Bembo, a copy of a roman cut
by Francesco Griffo for the Venetian printer Aldus Manutius,
which was first used in Cardinal Bembo's *De Aetna* in 1495.
Bembo was the forerunner of the standard European type
of the next two centuries and was the model followed by Garamond.

The first edition was printed in offset on 70# Finch Opaque Text
at The Stinehour Press in Lunenburg, Vermont,
and bound at Acme Bookbinding in Charlestown, Massachusetts.